DIRTY WORK

DIRTY

FARRAR, STRAUS AND GIROUX

NEW YORK

WORK

Essential Jobs
and the Hidden Toll of
Inequality in America

Eyal Press

Farrar, Straus and Giroux
120 Broadway, New York 10271

Copyright © 2021 by Eyal Press
All rights reserved
Printed in the United States of America
First edition, 2021

Portions of this book originally appeared, in different form, in *The New Yorker* and *The New York Times Magazine*.

Library of Congress Cataloging-in-Publication Data
Names: Press, Eyal, author.
Title: Dirty work : essential jobs and the hidden toll of inequality in America / Eyal Press.
Description: First edition. | New York : Farrar, Straus and Giroux, [2021] | Includes bibliographical references and index.
Identifiers: LCCN 2021011628 | ISBN 9780374140182 (hardcover)
Subjects: LCSH: Equality—United States. | Occupations—United States.
Classification: LCC HM821 .P668 2021 | DDC 331.700973—dc23
LC record available at https://lccn.loc.gov/2021011628

Designed by Janet Evans-Scanlon

Our books may be purchased in bulk for promotional, educational, or business use. Please contact your local bookseller or the Macmillan Corporate and Premium Sales Department at 1-800-221-7945, extension 5442, or by email at MacmillanSpecialMarkets@macmillan.com.

www.fsgbooks.com
www.twitter.com/fsgbooks • www.facebook.com/fsgbooks

10 9 8 7 6 5 4 3 2 1

Some of the names in this book have been changed.

In memory of Talma, my beloved aunt and a great lover of books

The powerless must do their own dirty work.
The powerful have it done for them.

—JAMES BALDWIN

Contents

DIRTY WORK

Introduction

One evening in May, in the city of Frankfurt, an American named Everett Hughes visited the home of a German architect. The year was 1948, and like much of the rest of Germany, Frankfurt lay in ruins. Crumbling villas lined the war-blasted boulevards, which the Allies had bombed repeatedly during the aerial campaign against the Nazis. Whole neighborhoods had been leveled to the ground. Driving around a few weeks earlier, Hughes and some companions had woven through the cratered streets of the decimated city center in search of a block whose storefronts and residential buildings had made it through the war unscathed. After a while, they gave up. "There was always at least one roof or house gone—more often half or more," he wrote in his diary.

Hughes had not come to Frankfurt to survey the wreckage. A sociologist at the University of Chicago, he was there to spend a semester teaching abroad. Born in 1897, he was a disciple of Robert Park, a former journalist and aide to Booker T. Washington who cofounded the Chicago school of sociology, which stressed the value of direct observation in the study of what Park called human ecology. A keen observer with a fondness for literature and a knack for seeing broad patterns in the details of small, seemingly singular events, Hughes rarely traveled far without a diary or journal in which he jotted down ideas that often made their way into his scholarly work.

In the journal he kept while in Frankfurt, Hughes described socializing with "liberal intellectual people who could be of any western country in their general ideas, attitudes and sophistication." The visit he paid to the architect was typical in this respect. They sat in a large studio filled with drawings, sipping tea and chatting about science, art, the theater. "If only the intelligent people of all countries could meet," a German schoolteacher who was also there remarked. At one point during the evening, after the schoolteacher complained that some of the American soldiers she'd encountered in Frankfurt (which was still under U.S. occupation) lacked manners, Hughes decided to bring up a more delicate subject. Was she aware, he asked, of the way that many German soldiers had comported themselves during the war?

"I am ashamed for my people whenever I think of it," the architect stated. "But we didn't know about it. We only learned all that later. And you must remember the pressure we were under; we had to join the party, we had to keep our mouths shut and do as we were told. It was a terrible pressure.

"Still, I am ashamed," the architect went on. "But you see, we had lost our colonies and our national honor was hurt. And these Nazis exploited that feeling. And the Jews, they *were* a problem . . . the lowest class of people, full of lice, dirty, poor, running about in their Ghettos in filthy caftans. And they came here and got rich by unbelievable methods after the first war. They occupied all the good places. Why, they were in the proportion of 10 to 1 in medicine and law and government posts."

At this point, the architect lost his train of thought. "Where was I?" he asked. Hughes reminded him that he had been complaining about how the Jews had "got hold of everything" before the war.

"Oh yes, that was it," the architect said. "Of course, that was no way to settle the Jewish problem. But there *was* a problem and it had to be settled some way."

Hughes left the architect's house shortly after midnight. But this conversation stayed with him. After returning to North America, he described it in a lecture at McGill University in Montreal. Fourteen years later, in 1962, a version of the lecture appeared in the journal *Social Problems*. By this point, numerous theories had emerged to explain the procession of horrors that had unfolded under the Nazis and culminated in genocide: the existence of a uniquely German "authoritarian personality"; the fanaticism of Adolf Hitler. Hughes focused on another factor that implicated people who were anything but fanatics and that was hardly unique to Germany. The perpetrators who carried out the ghastly crimes under Hitler were not acting solely at the behest of the führer, he argued. They were "agents" of "good people" like the architect who refrained from asking too many questions about the persecution of the Jews because, at some level, they were not entirely displeased.

"Holocaust," "Judeocide": various terms had been used to describe the Nazi campaign to exterminate the Jews. Hughes chose a more prosaic expression. He called it "dirty work," a term that connoted something foul and unpleasant but not wholly unappreciated by the more respectable elements in society. Ridding Germany of "inferior races" was not unwelcome even among educated people who were not committed Nazis, Hughes concluded from the architect's reflections on the "Jewish problem," variations of which surfaced in other conversations he had while in Frankfurt. "Having dissociated himself clearly from these people, and having declared them a problem, he apparently was willing to let someone else do to them the dirty work which he himself would not do, and for which he expressed shame," Hughes wrote of the architect. This was the nature of dirty work as Hughes conceived of it: unethical activity that was delegated to certain agents and then conveniently disavowed. Far from rogue actors, the perpetrators to whom this work was allotted had an "unconscious mandate" from society.

In recent years, a growing body of evidence has confirmed that the Nazis did manage to secure such a mandate. As the historian Robert Gellately shows in his 2001 book, *Backing Hitler*, the violent campaigns against Jews and other "undesirables" were hardly a secret to ordinary Germans, who knew about and not infrequently lent assistance to the drive for racial purification. In this sense, Hughes's article in *Social Problems*, titled "Good People and Dirty Work," was prescient. But as Hughes took pains to emphasize, he had not published his essay to establish this. "I do not revive the case of the Nazi *Endloesung* (final solution) of the Jewish problem in order to condemn the Germans," he wrote, "but to recall to our attention dangers which lurk in our midst always."

Raised in a small town in rural Ohio, Hughes had witnessed some of these dangers up close. He was the son of a Methodist minister whose commitment to racial tolerance won him no love from the Ku Klux Klan, which, one night, dispatched some of its white-robed emissaries to the Hughes household to burn a cross on the family lawn. The experience imbued Hughes with an awareness of the darker currents that ran through his own society and with a lifelong aversion to chauvinism of any kind. A skeptic who recoiled from the jingoism of the Cold War, Hughes had little patience for the notion that America was an exceptional nation immune to the moral lapses that befell other countries. After his essay on dirty work was published, the sociologist Arnold Rose wrote to *Social Problems* to complain that Hughes had understated the uniquely murderous nature of Nazi racial ideology. In response, Hughes emphasized, again, that he hadn't written it with the German experience foremost in mind. "[My essay] was addressed to North Americans . . . to put us—and especially the people of the U.S.A.—on guard against our own inner enemies," he affirmed. "We are so accustomed to racial violence and to violence of other kinds that we think little of it. That was the theme of my lecture in 1948. I repeat it more emphatically in 1963, when many of us Americans still practice private lynching, police torture, what amounts to inquisition and criminal trial by legislative bodies; and when the rest of us do not bother, dare, or have not found a way to stop it."

As the exchange suggests, Hughes was interested in raising questions about a dynamic that he was convinced existed in every society, not least his own. There was, to be sure, no moral equivalence between the injustices of postwar America and the atrocities of the Nazi era, which Hughes described as "the most colossal and dramatic piece of social dirty work the world has ever known." But less extreme forms of dirty work that took place in less autocratic countries still required the tacit consent of "good people." In fact, one could argue, this consent mattered far *more* in a democracy, where dissent was tolerated and public officials could be voted out of office, than in a dictatorship like Nazi Germany. Like their peers in other democratic countries, Americans had the freedom to question, and potentially stop, unethical activity that was carried out in their name.

"The question concerns what is done, who does it, and the nature of the mandate given by the rest of us to those who do it," wrote Hughes. "Perhaps we give them an unconscious mandate to go beyond anything we ourselves would care to do or even to acknowledge."

• • •

More than fifty years after Hughes's essay was published, the questions he posed bear revisiting. What kind of dirty work takes place in contemporary America? How much of this work has an unconscious mandate from society? How many "good people" prefer not to know too much about what is being done in their name? And how much easier is this to achieve when what gets done can be delegated to a separate, largely invisible class of "dirty workers"?

Since the winter of 2020, our collective reliance on invisible workers who help keep society running has been glaringly exposed. It came to light during the coronavirus pandemic, which prompted governors to issue lockdown orders and led tens of millions of jobs to disappear or be put on hold. The pandemic revealed the degree to which more privileged Americans with the luxury to work from home were dependent on millions of low-wage workers—supermarket cashiers, delivery drivers, warehouse handlers—whose jobs were deemed too critical to be halted. These jobs were often reserved for women and people of color, hourly workers toiling in the shadows of a global economy whose rewards had long eluded them. During the pandemic, the functions these laborers performed received a new designation: "essential work." This designation did little to alter the fact that many workers continued to be denied access to health care, paid sick leave, and, even as they risked exposure to a potentially fatal virus, personal protective equipment. Yet it underscored a basic truth, which is that society could not function without them.

But there is another kind of unseen labor that is necessary to society, work that many people see as morally compromised and that is even more hidden from view. The job of running the psychiatric wards in America's jails and prisons, for example, which have displaced hospitals as the largest mental health institutions in many states, resulting in untold cruelty and in routine violations of medical ethics among staff who acquiesce when security guards abuse incarcerated people. Or the job of carrying out "targeted killings" in America's never-ending wars, which have faded from the headlines even as the number of lethal strikes conducted with little oversight has steadily increased.

Critics of mass incarceration or of targeted drone assassinations would likely argue that this work is anything but essential. Yet it is necessary to the prevailing social order, solving various "problems" that many Americans want taken care of but don't want to have to think too much about, much less handle themselves. "Problems" like where to put all the people with severe mental illnesses who lack access to care in their communities, hundreds of thousands of whom have been warehoused in jails and prisons and quickly forgotten about. Or how to continue fighting endless wars when the nation has lost its appetite for expensive foreign interventions and uncomfortable debates about torture and indefinite detention, a predicament that the use of armed drones resolved.

In the past year, some of the workers who take care of such problems became slightly less invisible—most notably, the predominantly Black and brown workers who man the "kill floors" of America's slaughterhouses, where animals are hacked apart under brutal conditions that consumers never see in order to satisfy the popular demand for cheap meat. The coronavirus pandemic drew attention to the physical risks endured by the line workers in beef, pork, and chicken plants, which were ordered to stay open even as scores of laborers died and tens of thousands fell ill. Slaughterhouse workers, like many dirty workers, are often exposed to extreme physical risks on the job, a product of the harsh conditions in their industries and of their relative powerlessness. But they are even more susceptible to another, less familiar set of occupational hazards, owing to the unpalatable nature of the jobs they do. In the eyes of many Americans, the mass killing of animals in industrial slaughterhouses, like the mass confinement of mentally ill people in jails and prisons, evokes discomfort, even disgust and shame. These feelings inevitably color how, to the extent that they are noticed by the public, the workers who do the killing and confining are perceived—and, to some extent, how these workers see themselves. In their classic book, *The Hidden Injuries of Class*, the sociologists Richard Sennett and Jonathan Cobb called for shifting the focus of class analysis away from material conditions to "the moral burdens and the emotional hardships" that workers bear. For dirty workers, these burdens include stigma, self-reproach, corroded dignity, shattered self-esteem. In some cases, they include post-traumatic stress disorder and "moral injury," a term that military psychologists have used to describe the suffering that some soldiers endure after they carry out orders that transgress the values at the core of their identity.

The idea that work can be morally injurious has not gone entirely unnoticed. At the height of the coronavirus pandemic, it was described in often-moving detail in articles about physicians and nurses who were forced to make excruciating decisions—which patients should be hooked up to ventilators? who should be kept alive?—as hospitals were inundated with COVID-19 cases. "None of us will ever be the same," wrote an ER doctor in New York City who worked on the front lines of the pandemic and published a firsthand account of the anguish that she and her colleagues felt. Notably, though, it took an unforeseen crisis to thrust doctors into such a role, a crisis that eventually abated. In the case of many dirty workers, the wrenching choices—and the anguish they can cause—occur on a daily basis because of how society is organized and what their jobs entail. Unlike doctors, moreover, these workers are not lionized by their fellow citizens for working in a profession that is widely viewed as noble. To the contrary, they are stigmatized and shamed for doing low-status jobs of last resort.

People who are willing to do morally suspect things simply to earn a paycheck deserve to be shamed, some may contend. This is how many advocates of migrant rights feel about the Border Patrol agents who have enforced America's inhumane immigration policies in recent years. It is why some peace activists have accused drone operators involved in targeted killings of having blood on their hands. These activists have a point. The dirty workers whose stories unfold in the pages that follow are not the primary victims of the systems in which they serve. To the people on the receiving end of their actions, they are not victims at all. They are perpetrators, carrying out functions that often cause immense suffering and harm.

But pinning the blame for dirty work solely on the people tasked with carrying it out can be a useful way to obscure the power dynamics and the layers of complicity that perpetuate their conduct. It can also deflect attention from the structural disadvantages that shape *who* ends up doing this work. Although there is no shortage of it to go around, the dirty work in America is not randomly distributed. As we shall see, it falls disproportionately to people with fewer choices and opportunities—high school graduates from depressed rural areas, undocumented immigrants, women, and people of color. Like jobs that pay poorly and are physically dangerous, such work is chiefly reserved for less privileged people who lack the skills and credentials, and the social mobility and power, that wealthier, more educated citizens possess.

The dilemmas and experiences of these workers tell a larger story about contemporary America, illuminating a dimension of inequality that has escaped the notice of economists. The concentration of wealth in fewer and fewer hands, the stagnation of median wages: this is how inequality is typically measured and described, through statistics that dramatize how few Americans have benefited from the economic growth of recent decades. The statistics are indeed dramatic. According to the economists Thomas Piketty, Emmanuel Saez, and Gabriel Zucman, the share of national income going to the top 1 percent of Americans nearly doubled between 1980 and 2014, while the share going to the bottom half fell by nearly 50 percent. According to another study, the four hundred richest Americans now command more wealth than all African Americans combined.

But economic inequality mirrors and reinforces something else: moral inequality. Just as the rich and the poor have come to inhabit starkly different worlds, an equally stark gap separates the people who perform the most thankless, ethically troubling jobs in America and those who are exempt from these activities. Like so much else in a society that has grown more and more unequal, the burden of dirtying one's hands—and the benefit of having a clean conscience—are increasingly functions of privilege: of the capacity to distance oneself from the isolated places where dirty work is performed while leaving the sordid details to others. People with fewer advantages are not only more likely to do this work; they are also more likely to be faulted for it, singled out as "bad apples" who can be blamed when systemic violence that has long been tolerated and perhaps even encouraged by superiors occasionally comes to light. Politicians and the media often treat these moments of exposure as "scandals" and focus on the corrupt individuals involved, a display of outrage that can end up hiding the more mundane injustices happening every day. Meanwhile, the higher-ups, and the "good people" who have tacitly condoned what they are doing, remain untarnished, free to claim that they knew nothing about it while casting judgment on the scapegoats who were singled out.

To be sure, not all dirty workers see what they are doing as compromised. Some derive satisfaction from their jobs. There is, moreover, an argument to be made that a certain amount of dirty work is inevitable in any society and that plenty of elite white-collar professionals—Wall Street bankers who sell shady financial products, software engineers who design hidden tracking mechanisms that enable companies to collect users' personal data without their knowledge—do jobs that are morally suspect. But for these elites, there are significant upsides: in the case of Wall Street bankers, lavish salaries and bonuses; in the case of software engineers, a place in the upper tiers of the meritocracy. In a society where worldly success has long been perceived as a mark of good character, accomplishing such feats has a positive moral valence, conferring virtue on the winners who have risen to the top of the social order. Successful meritocrats may also feel more emboldened to complain or resign when pressed to do something ethically compromising. Exercising such options is never free from risk, but it is invariably easier for people with the skills and credentials to land other desirable jobs.

The dirty workers featured in this book lack this luxury. Most feel trapped in what they are doing, clinging to their jobs in order to make ends meet and because they have no better options. Not all of these workers are poor. For some, dirty work may indeed offer a path out of poverty. In some cases, it is the one job within reach that comes with health benefits or that pays slightly more than the minimum wage. But the benefits or higher wages come at a steep price: the cost of feeling degraded and defiled, of dirtying one's hands in a disreputable job that others look down on. Insofar as their livelihoods depend on such jobs, dirty workers are doubly burdened, experiencing economic precarity while simultaneously bearing the psychic toll of doing morally treacherous work.

The familiar, colloquial meaning of "dirty work" is a thankless or unpleasant task. In this book, the term refers to something different and more specific. First, it is work that causes substantial harm either to other people or to nonhuman animals and the environment, often through the infliction of violence. Second, it entails doing something that "good people"—the respectable members of society—see as dirty and morally compromised. Third, it is work that is injurious to the people who do it, leading them either to feel devalued and stigmatized by others or to feel that they have betrayed their own core values and beliefs. Last and most important, it is contingent on a tacit mandate from the "good people," who see this work as a necessary part of the social order but don't explicitly assent to it and can, if need be, disavow responsibility for it. For this to be possible, the work must be delegated to *other* people, which is why the mandate rests on an understanding that someone else will handle the day-to-day drudgery.

This book does not present a comprehensive survey of all the jobs that share these features. What it offers instead is a series of case studies that illuminate the dynamics of dirty work in different areas of American life. Part 1 examines the dirty work that takes place inside the mental health wards of the nation's prisons, where chilling abuses are perpetrated on a regular basis. While these abuses are easy to blame on low-ranking guards who behave sadistically, the custodians of our jails and prisons are the agents of a society that has criminalized mental illness, making brutality and violence all but inevitable. Part 2 examines another kind of violence, carried out from a distance by imagery analysts who help select targets for lethal drone strikes. Public officials have often portrayed these strikes as "pinpoint" and "surgical"—that is, as the opposite of dirty. As we'll see, the reality for many of the "virtual warriors" involved is more disturbing, suggesting that distance and technology can make warfare and violence more rather than less morally troubling. Like prison workers, the combatants in the drone program perform government functions, executing policies that have the presumed backing of public officials and many citizens. But dirty work can also take place in institutions with no formal connection to the state—industrial slaughterhouses, for example, where part 3 of the book is set. The workers in these slaughterhouses are our agents as well, not because they carry out public functions, but because they cater to our consumption habits. The lifestyles of many Americans—the food we eat, the cars we drive—are sustained by dirty work. In the final section of the book, I explore how this is true not only in America but in much of the world, examining the dirty work behind the lubricants of global capitalism: the fossil fuels that are drilled and fracked by dirty workers in places like the Gulf of Mexico; the cobalt that is mined in Africa before making its way into the wireless devices that have propelled the digital revolution.

One characteristic common to nearly all forms of dirty work is that they are hidden, making it easier for "good people" to avoid seeing or thinking about them. The desire not to witness things that are filthy or repugnant is hardly new. "Dirtiness of any kind seems to us incompatible with civilization," observed Sigmund Freud in *Civilization and Its Discontents*. "Indeed, we are not surprised by the idea of setting up the use of soap as an actual yardstick of civilization." Among the thinkers Freud influenced was the German social theorist Norbert Elias, whose best-known work, a two-volume study titled *The Civilizing Process*, traced the evolution of morals and manners in the West, showing how behavior that came to be seen as disturbing or distasteful (spitting, displaying violence and aggression) was gradually removed from public life. Elias completed his book in 1939, which may explain why for several decades it was ignored: in the shadow of Nazism, it seemed to many that the savage face of Western civilization had been unmasked. But Elias did not equate the "civilizing process" with moral progress. Like Freud, he linked it to rising social inhibitions, which led to practices regarded as unseemly being carried out more discreetly. In theory, this could make objectionable practices more rather than less pervasive. "The distasteful is *removed behind the scenes of social life*," Elias observed. "It will be seen again and again how characteristic of the whole process that we call civilization is this movement of segregation, this hiding 'behind the scenes' of what has become distasteful."

Behind the scenes is where the dirty work in America unfolds, in the chambers and recesses of remote institutions such as prisons and industrial slaughterhouses—institutions that tend to be located in isolated areas with a high concentration of poor people and people of color. The workers who toil in these zoned-off worlds are, in a sense, America's "untouchables," performing morally tainted jobs that society depends on and tacitly condones but that have been rendered invisible. The invisibility is sustained with physical barriers—fences and walls that cordon off the places where dirty work takes place. It is reinforced by legal barriers—secrecy laws that limit what the public is permitted to know. But perhaps the most important barriers are the ones in our own minds, mental filters that block out uncomfortable realizations about the things we are willing to countenance.

In the margins of the journal he kept while in Frankfurt, Everett Hughes jotted down a phrase for people who erected such barriers. He called them "passive democrats." Passive democrats were people with seemingly enlightened attitudes "who don't mean ever to do anything about anything, except carry on delightful, disinterested conversation." The problem with such people was not that they didn't know about the unconscionable things going on around them. It was that they lacked what Hughes called "the *will* to know." To maintain a clean conscience, they preferred to be kept in the dark.

It's hard to say how much of a difference it would have made if the passive democrats in Nazi Germany had been more active; they lived in a dictatorship, after all, where dissent was crushed and the state demanded absolute obedience from its subjects. But as noted, Hughes wasn't thinking primarily of Nazi Germany when he wrote his essay about dirty work. He was thinking of his fellow Americans, citizens of a democracy in which active engagement could make a difference, stirring debate about whether morally objectionable practices should go on.

In the decades since Hughes's essay appeared, the passivity of Americans seems only to have deepened. In recent presidential elections, tens of millions of voters have not bothered to exercise a right for which prior generations fought and died. Thanks to technology, information has never been easier for ordinary people to access. It has also never been easier to avert one's eyes by clicking on another link when something disturbing comes up. In a culture of distraction and diminishing attention spans, who has the patience to wade through troubling revelations that might make one feel implicated in some way? Or to feel the flicker of conscience for long enough while surfing the internet to remember the experience the next day? Studies of college graduates have documented a decline in empathy in recent years. Along with the will to know, the will to imagine what it is like to stand in someone else's shoes appears to be slackening.

A nation of passive democrats is a nation where disturbing practices can flourish without too many questions asked. This is unfortunate, for a great deal can be learned about the moral condition of our society by tracing the threads of dirty work through the fabric of American life. As we'll see, we are all entangled in these threads, even if they are imperceptible to us. The philosopher Charles Mills has argued that the advantages accorded to whites in Western societies are enshrined in an invisible "racial contract," an implicit agreement that nonwhites are "subpersons" that governs the racial order, even though it is unnoticed and unacknowledged by many of its beneficiaries. An invisible contract governs dirty work as well, the terms of which ensure that those who tolerate and benefit from it don't have to know too much about it. Like the racial contract, this arrangement is not spelled out in any formal document, which makes it easy to ignore and, when it is noticed or brought up, equally easy to blame on others or attribute to large external forces that cannot be changed. This is a mistake. As immutable as it may seem, the dirty work in America is not foreordained. It is a product of specific decisions made by real people that could, in theory, be unmade: policies that were enacted; laws that were put in place; decisions that were reached about everything from how to fight our wars to where to confine some of our most vulnerable fellow citizens. How we think about this work reveals something fundamental about our society—our values, the social order we unconsciously mandate, and what we are willing to have done in our name.

PART I

BEHIND THE WALLS

1

Dual Loyalties

Shortly after Harriet Krzykowski began working at the Dade Correctional Institution in Florida, a prisoner whispered to her, "You know they starve us, right?" It was the fall of 2010, and Harriet, a mental health counselor, had been hired by Dade, a state prison roughly forty miles south of Miami, to help incarcerated people with clinical behavioral problems follow their treatment plans. The prisoner was housed in the facility's mental health ward, known as the Transitional Care Unit, a cluster of two-story buildings connected by breezeways and equipped with one-way mirrors and surveillance cameras. At first, Harriet assumed he was just imagining things. "I thought, oh, this guy must be paranoid or schizophrenic," she said. Then she heard a prisoner in another wing of the TCU complain that meal trays often arrived at his cell without food. After noticing that several of the men in the TCU were alarmingly thin, she decided to discuss the matter with Dr. Cristina Perez, who oversaw the inpatient unit.

At the time, Harriet was thirty years old. She had pale skin, blue eyes, and an air of shy reserve. The field of correctional psychology attracted its share of idealists who tended to see all prisoners as society's victims and who distrusted anyone wearing a security badge; corrections officers called such people "hug-a-thugs." The label did not fit Harriet, who had never worked in a correctional facility before and who arrived at Dade acutely aware of the risks of her new job. There were, she knew, rapists, pedophiles, and murderers at the prison, convicted felons who inspired fear in her, not pity. The guards at Dade performed a difficult job that merited respect, she believed, not least for watching the backs of less experienced employees who did not wear security badges. If any corrections officers behaved improperly, she assumed her superiors would want to know about it.

Dr. Perez was in her forties and had an aloof, unruffled manner. When Harriet told her that she'd heard "guys aren't getting fed," she did not seem especially concerned. "You can't trust what inmates say," she reminded Harriet. When Harriet informed her that the complaints were coming from disparate wings of the TCU, Dr. Perez assured her that this was not unusual, because prisoners often devised innovative methods to "kite" messages across the facility.

Harriet mentioned that she had overheard some security guards heckling prisoners. "Go ahead and kill yourself—no one will miss you," an officer told one of them in her presence. Again, Dr. Perez seemed unfazed. "It's just words," she said. Then she leaned forward and imparted some advice: "You have to remember that we have to have a good working relationship with security."

Not long after this conversation, Harriet was working a Sunday shift when a guard told her that because of a staff shortage the prisoners in the TCU would not be allowed into the facility's recreation yard. The yard was a cement quadrangle with weeds sprouting through the cracks and little in the way of amenities, but for many people in the TCU it was the only place to get some fresh air and exercise. Overseeing this activity was among Harriet's weekend responsibilities. The following Sunday, access to the recreation yard was again denied. The closures continued for weeks, always with a new explanation that sounded to Harriet increasingly like a pretext. When she eventually pressed a security officer on the matter, he told her, "It's God's day, and we're resting." In an email to Dr. Perez, Harriet relayed this exchange and indicated her frustration.

A few days later, Harriet was running a "psycho-educational group"—an hour-long session in which prisoners gathered to talk while she observed their mood and affect. After a dozen participants filed in, she peered up and noticed that the guard who had been standing by the door had wandered off. She was on her own, in a roomful of prisoners. Harriet completed the session without incident; afterward, she assumed the officer must have been summoned to deal with an emergency. But later, when she was in the rec yard, the guard on duty vanished as well, once again leaving her unprotected amid a group of prisoners.

Around the same time, the metal doors that corrections officers controlled to regulate the traffic flowing through the different units of the TCU started opening more slowly for Harriet. Not infrequently, several minutes passed before an officer in one of the security bubbles buzzed her through, even when she was the only staff member in a hallway full of prisoners. Harriet tried not to appear flustered when this happened, but, she later recalled, "it scared the hell out of me."

In theory, the TCU was designed to provide mentally ill prisoners with a safe environment in which they would receive treatment before returning to the main compound at Dade. In reality, Harriet noticed, many of the people in her care were locked up in single-person cells for months, rarely interacting with anyone. Solitary confinement was supposed to be reserved for individuals who had committed serious disciplinary infractions. In forced isolation, the men in the TCU often deteriorated rapidly, appearing haggard, stricken, vacant-eyed. "So many guys would be mobile and interactive when they first came in, and then a few months later they would be sleeping in their cells in their own waste," Harriet said.

Despite her inexperience, Harriet was coming to doubt whether the TCU was fulfilling its stated mission. She was also convinced the guards at Dade were punishing her for the email she had sent to Dr. Perez about her difficulty accessing the rec yard. But complaining about the situation would only lead to more retaliation, she feared. She didn't even tell her husband, Steven, out of concern that if she voiced her misgivings, he would insist that she give notice, further unsettling their precarious financial situation.

At the time, Harriet and her husband lived at her mother's house, in Miami, with their two young children. He was an unemployed computer-systems engineer. She was earning twelve dollars an hour at Dade. To get by, they supplemented her modest paycheck with food stamps and occasional loans from her mother. The experience of hardship was familiar to Harriet. Born in a small town in northwestern Missouri, she was seven years old when her mother drove her and her older sister to a battered women's shelter to escape the furies of her father, a heavy drinker who precipitated their departure by hurling the family cat against the wall. Five years later, after her parents divorced, Harriet found herself in an even smaller town in Illinois, where her mother's dream of making a living as an artist (she was a potter) quickly ran aground. They survived instead on public assistance and a job her mother found as a clerk at a gas station, living in a modest house where the cupboards were often bare.

After Harriet graduated from high school in 1998, she and her mother moved to Miami, where their financial situation improved. Her mother became a nurse, and Harriet enrolled at Florida International University. She decided to major in psychology, captivated by the idea of helping people rein in their destructive impulses so they could lead more fulfilling lives. For a while, though, Harriet struggled to control her own destructive impulses. She binge-ate and put on fifty pounds, then starved herself and lost sixty-five. She did a lot of clubbing and a lot of drugs, adjusting to the culture shock of Miami by plunging into its hedonistic nightlife. The experience was exhilarating at first, but the thrill eventually wore off. After one of her roommates burned through all his savings to service a raging drug habit, Harriet quit the party scene and spent a year focusing on taking better care of herself. One night, she had a dream about a childhood friend named Steven, whom she'd met at a bus stop in fifth grade. She got in touch with him, inviting him to visit. He showed up a few weeks later and ended up sticking around. In 2007, they exchanged wedding vows.

By then, Harriet had earned a bachelor's degree in psychology and enrolled in a master's program in mental health counseling. She wanted to become a forensic psychologist. But Florida was in deep recession, racked by the collapse of the housing bubble and the 2008 financial crisis, which devastated an economy heavily dependent on real estate speculation. Harriet had no luck finding work until she spotted a listing posted by Corizon, the private contractor in charge of providing mental health services at Dade.

Even at the height of the economic crisis, jobs in corrections were plentiful in Florida; the state had the third-largest prison population in the country, behind only Texas and California. Ensuring that these prisoners received psychiatric care was not a matter of choice. It was a constitutional obligation, thanks to *Estelle v. Gamble*, a 1976 case in which the Supreme Court held that "deliberate indifference to serious medical needs of prisoners" amounted to cruel and unusual punishment.

Around the same time, in another case, *O'Connor v. Donaldson*, the Supreme Court ruled that a Florida man named Kenneth Donaldson had been confined for fifteen years in a state psychiatric hospital against his will, violating his constitutional rights. The ruling added momentum to the campaign to "deinstitutionalize" the mentally ill, whose plight was dramatized in a series of journalistic exposés that drew attention to the dismal conditions in the nation's mental health hospitals, where catatonic patients in dingy rags were crammed into bleak, filth-strewn wards. Activists in the disability rights movement decried the existence of asylums filled with "naked humans herded like cattle," as one account put it. In the decades that followed, states across the country began shutting down these facilities, both to save money and to appease advocates demanding reform.

The reformers' vision was a noble one, with origins that could be traced back to the 1963 Community Mental Health Act, signed by John F. Kennedy, who proposed creating a network of fifteen hundred community mental health centers so that the "cold mercy of custodial care would be replaced by the open warmth of community." But fulfilling this vision proved more challenging, not because it was inherently flawed, but because of choices that politicians and the people who elected them made. As the sociologist Christopher Jencks noted in his book *The Homeless*, the money saved by closing state mental hospitals could have been used to fund affordable housing and outpatient services. Unfortunately for advocates of reform and deinstitutionalization, the closures coincided with a new climate of fiscal austerity and conservative tax revolts. In Washington, the Reagan administration tightened eligibility for federal disability benefits, pushing more than a million people off the rolls. Meanwhile, state legislators pressed hospitals to discharge even chronically ill patients, who soon began inundating the streets. By 1987, there were 100,000 mentally ill homeless people and 1.7 million others who were too sick to hold down jobs, a situation without parallel in other Western societies. "No other affluent country has abandoned its mentally ill to this extent," Jencks observed.

Deinstitutionalization coincided with another trend unique to America: a wave of punitive criminal justice measures—mandatory minimum sentences, truth-in-sentencing laws—that fueled an unprecedented expansion of the prison population. Mass incarceration would have a particularly devastating effect on African Americans, who were arrested, convicted, and imprisoned at vastly higher rates than whites. It would have equally devastating consequences for the mentally ill, who were ensnared in the criminal justice system at similarly alarming rates. In the decades to come, as many as one in two people with severe mental illnesses would get arrested in some parts of the country, often for minor offenses that were a direct result of their conditions. Instead of the open warmth of community, many were subjected to an even colder form of custodial care, in the nation's growing archipelago of jails and prisons.

By the 1990s, prisons were becoming America's new asylums, warehousing more and more people in dire need of psychiatric care. The situation was particularly extreme in Florida, which spent less money per capita on mental health than any other state except Idaho. Meanwhile, between 1996 and 2014, the number of Florida prisoners with mental disabilities grew by 153 percent, three times faster than the overall prison population.

In its *Estelle* decision, the Supreme Court determined that neglecting to furnish incarcerated people with psychiatric care violated their rights and was unacceptable. But the Court failed to clarify *how* such care could be delivered in a punitive environment where the paramount concern was security. According to medical ethicists, correctional psychiatrists often felt a "dual loyalty"—a tension between the impulse to defer to corrections officers and the duty to care for their clients. Because guards provided staff with crucial protection, it could be risky to defy them. Yet if mental health professionals were too acquiescent, they risked becoming complicit in practices that caused their patients grievous harm.

"DON'T BE A WITNESS"

After Harriet met with Dr. Perez, she told herself, "Maybe I'm being too sensitive—boys will be boys." Aware that she was a newcomer to the world of prisons, she decided that the guards at Dade were far more qualified than she was to determine how to maintain order.

At a morning staff meeting, a psychotherapist named George Mallinck-rodt aired a different view. The day before, Mallinckrodt announced, a patient had shown him a collage of grisly bruises on his chest and back. The injuries had been sustained when a group of guards handcuffed him, dragged him into a narrow hallway, and stomped on him over and over again with their boots. Several other prisoners had corroborated this account, Mallinckrodt told his colleagues. At the meeting, he accused security of "sabotaging our caseload" and called for immediate action to be taken.

Harriet did not attend this meeting, but she heard about Mallinck-rodt's allegations and, like the rest of her peers, dismissed them as exaggerated. "I thought 'sabotaging' was a pretty strong word, a loaded word," she said. Mallinckrodt was known to be on chummy terms with some of the men in the TCU—too chummy, she believed. "I thought, you're just too enmeshed in your clients, you can't see things objectively," she said. "I thought he'd become an advocate. You know, a 'hug-a-thug.'"

Harriet had no intention of playing such a role. She just wanted to do her job. Yet she soon discovered that her power to make even basic decisions was extremely circumscribed. When she'd been hired, for example, Dr. Perez had told her that in addition to assisting with individual treatment plans, she would be responsible for helping people in the TCU participate in twenty hours of activity a week, as mandated by state law. Yet every time Harriet proposed something—music therapy, yoga—her superiors rejected the idea. Invariably, the reason cited was that it posed a risk to security, even when the activities were designed to alleviate aggression. One day, Harriet brought in a box of chalk, in the hope that prisoners could draw on the pavement in the rec yard. On another occasion, she gave a red rubber ball to a man who had schizophrenia, thinking that he would benefit from tactile play. An officer returned both items to her, ostensibly because they posed safety hazards. Eventually, Harriet came to feel that she was being taught a lesson about knowing her place. "I kept getting the message that mental health doesn't override security," she said. "Whatever security says goes."

The restrictions frustrated Harriet, but she knew from experience how dangerous it could be to alienate security and work without protection, as had happened after the email she'd sent to Dr. Perez about the rec yard.

One day, Harriet was in the yard when the guard on duty told her that he needed to step away. "I'll be back," he said. A short while later, a prisoner sidled up to her and put his hands on her backside. Harriet thought of screaming for help but feared further provoking the man, who was extremely psychotic. Instead, she froze. After a moment, she slipped away without looking back. The prisoner didn't follow her. Even so, the incident left her deeply shaken. "He definitely could have overpowered me," she said. "I could have been assaulted, raped—anything."

Navigating such dangers would have been stressful enough in a tidy, well-run work environment. The setting in which Harriet worked was filthy and run-down—the walls coated in mildew, the hallways caked in grime. Cells in the TCU flooded and weren't mopped up for days, creating appalling smells; ceilings routinely sprang leaks. In the staff break room, cockroaches had the run of the kitchen area, infesting even the microwave. To avoid using it, Harriet started making ramen noodles for lunch, which she prepared with water poured directly from the kitchen faucet (it was so hot that she didn't have to boil or microwave it).

One Saturday in June 2012, Harriet was finishing a shift when she heard that a prisoner in the TCU named Darren Rainey had defecated in his cell and was refusing to clean it up. Rainey was fifty years old, a tall, broad-shouldered man who sometimes gave people unnerving looks, Harriet later recalled, "as though he was trying to see inside you." He had been convicted of cocaine possession and suffered from severe schizophrenia.

"What's going on with Rainey?" Harriet asked a guard.

"Oh, don't worry, we'll put him in the shower," he told her.

Harriet heard this and felt reassured. "I was thinking, okay, lots of times people feel good after a shower, so maybe he will calm down. A nice, gentle shower with warm water."

The next day, Harriet returned to work and learned from some nurses that a couple of guards had indeed escorted Rainey to the shower the previous night. She also learned that he hadn't made it back to his cell. He had collapsed in the stall while the water was running. At 10:07 p.m., he was pronounced dead.

Harriet assumed that Rainey must have had a heart attack or somehow died by suicide. The nurses told her this was not what had happened, explaining that he had been deliberately locked in the shower by guards, who directed a stream of scalding water at him through a hose they controlled from the outside. The water flowing through the rigged-up hose was 180 degrees, hot enough to brew a cup of tea—or, it soon occurred to Harriet, to steam a bowl of ramen noodles. It would later be revealed that Rainey had burns on 90 percent of his body and that his skin fell off at the touch.

Harriet was stunned. Surely, she told the nurses, the incident would prompt a criminal investigation.

"No," one of them told her. "They're gonna cover this up."

In the days after Rainey's death, Harriet heard from several prisoners in the TCU that Rainey was not the first person who had been locked in that shower. He was only the first person to die there. Earlier she would have rolled her eyes at such a claim. Now she wondered how she could have been so blind. Yet as distraught as she was, she did not file a report calling for the guards who killed Rainey to be held accountable. Neither did anyone else on the mental health staff. "I thought, *somebody* has to report it, and it has got to come from the inside, but it's not going to be me," she said. One reason was her memory of the backlash she'd provoked for reporting something incredibly minor by comparison. Another was her fear that any employee who became too vocal would end up getting fired.

This was a legitimate concern. A year earlier, after George Mallinckrodt heard about the prisoner who'd been stomped on, he consulted the website of the Florida Department of Corrections. It stated that any employee who suspected abuse was obligated to report it. Mallinckrodt had first learned about the attack from a fellow counselor who had called him after she had witnessed it. Choking back tears, the counselor said that she had seen a group of guards batter the prisoner in a hallway, slamming their boots into his rib cage as he flailed on the ground. She had watched the assault through a window that faced onto the corridor, which had no cameras. The guards, she said, stopped pummeling the man only when she started yelling at them. (All of these details matched what Mallinckrodt subsequently heard from the participants in his group therapy session, where the victim lifted up his shirt to reveal his bruises.)

The counselor who relayed this account to Mallinckrodt had attended the staff meeting where he had spoken out, but she had remained silent. As incensed as she was, she did not intend to report anything, out of fear that the guards would turn on her if she dared to cross them. Mallinck-rodt's other peers also did not respond to his call for action.

Because no one else was willing to come forward, Mallinckrodt decided to act on his own. In July 2011, he filed an incident report with both the Florida Department of Corrections and the Florida inspector general's office in Tallahassee. Around this time, a new warden, Jerry Cummings, arrived at Dade. Mallinckrodt arranged a meeting, at which he told Cummings about the beating and other abuses that his patients had described to him, including incidents of security officers' taunting mentally ill prisoners until they screamed, banged their heads against the walls, or defecated in their cells—which, in turn, triggered more brutality from the guards. According to Mallinckrodt, Cummings seemed concerned and sympathetic, leaving him optimistic that changes would occur.

Shortly thereafter, a change did occur, albeit not the one he was anticipating. One afternoon, an officer stopped Mallinckrodt at the entrance gate to the prison as he was returning from a lunch break. He had just been fired, the officer informed him, on the grounds that his lunch breaks had been stretching on for hours. Mallinckrodt did not deny that he took long lunches, but plenty of staff members did this. He was the only one who had started raising his voice about abuse.

Harriet did not want to end up in a similar situation. "We couldn't have afforded it," she said. So she refrained from raising her voice, a decision that had other costs. At lunch, she rarely ate anything because, after a few hours in the TCU, she would feel queasy and lose her appetite. She also started losing her hair, which began to fall out in mysteriously large clumps. At first, Harriet thought she might be suffering from an iron deficiency. Eventually, it dawned on her that the cause was emotional stress. Embarrassed that her scalp was showing, she took to wrapping scarves around her head to cover up the bald spots.

In addition to these symptoms, Harriet began to feel something else: a sense of powerlessness that recalled the darkest moments of her childhood, when she had witnessed the erratic behavior of her father and felt powerless to stop it. Her sister had occasionally stood up to his drunken tirades, whereas Harriet had always tried to win his affection by impressing him. When this failed, as it inevitably did, she retreated into herself. Now, once again, she was trapped in an environment where she was too afraid to speak out. Even observing misconduct in the TCU was risky, because the guards were on alert for anyone who might expose them. If abuse was happening, "the politically safest thing was to excuse yourself and go to the bathroom," she said. "Don't be a witness. Just do your job and go home."

In 2013, Harriet was promoted to staff counselor, at which point she began providing individual therapy for prisoners. One day, a patient of hers named Harold Hempstead told her that he had become preoccupied with the murder of Darren Rainey. A wiry man with hazel eyes, Hempstead, who had been convicted of burglary, occupied the cell directly below the shower where Rainey had been tortured. That night, he heard Rainey scream repeatedly, "Please take me out! I can't take it anymore!" He also heard him kick at the stall door. Eventually, there was a heavy thud—the sound of Rainey collapsing to the ground, Hempstead later surmised—followed by the voices of guards calling for medical help. A short while later, Hempstead watched as a gurney with Rainey's naked body on it was wheeled past his cell.

Hempstead kept a diary in which he had recorded the names of four other people who had been subjected to what he called the "shower treatment." He had even noted the dimensions of the stall. In the weeks after Rainey's death, Hempstead told several mental health counselors in the TCU that he felt haunted by what he had heard and seen. They warned him that if he told them too much, they would have to write an incident report that would be forwarded to security officials, exposing him—and, by implication, them—to retaliation. Around this time, two of the guards who took Rainey to the shower, including a former football player named Roland Clarke, were promoted. (Both officers later resigned; their files included no indication of wrongdoing.) But Hempstead, who had been diagnosed with obsessive-compulsive disorder, refused to let the matter drop. He told Harriet that he had begun filing grievance reports about Rainey's murder, in the hope of triggering an investigation.

Harriet wasn't sure if Hempstead, who professed to be a devout Christian, was motivated by compassion for Rainey or by a less high-minded impulse—a desire to embarrass the guards at Dade or to leverage a transfer out of the TCU. In the end, however, she encouraged him. "I thought that, therapeutically, writing it all down would be good for him," she said. This advice was consistent with her general approach to her job, which was to try to make small differences in the lives of the people under her care while ignoring the problems that she lacked the power to alter. Harriet was the only mental health counselor at Dade who did not pressure Hempstead to remain silent about Rainey's death, and he was grateful to her. But when he asked her whether she would back him up if he succeeded in drawing attention to the abuses at Dade, she hesitated. "I said, 'Well, yeah.' But I didn't honestly know if I would honor that."

"A COLLECTIVE PUBLIC YAWN"

On May 17, 2014, Julie Brown, a reporter at the *Miami Herald*, published an article in the paper about the abuse of mentally ill prisoners in the TCU at Dade. Below the headline was a photograph of Darren Rainey, clad in prison blues. Brown's main source was Harold Hempstead, who had turned over copies of the complaints that Harriet had encouraged him to write. The article indicated that after being interviewed, Hempstead had been threatened with solitary confinement and other forms of punishment by three corrections officers.

After the *Herald* article appeared, Jerry Cummings, the warden at Dade, was placed on administrative leave, and some prisoner rights advocates questioned whether the Department of Corrections had tried to cover up a murder. Few questioned why the burden of exposing it had fallen to a prisoner rather than to one of the prison's mental health or medical professionals. The duty to protect patients from harm is a core principle of medical ethics. According to the National Commission on Correctional Health Care, an offshoot of the American Medical Association that issues standards of care for prisons, any mental health professional who is aware of abuse is obligated "to report this activity to the appropriate authorities."

But counselors are not likely to fulfill this duty if they have reason to fear that doing so will put them in danger, a feeling that is pervasive. In a 2015 survey by the Bureau of Correctional Health Services in New York City, more than one-third of mental health personnel working in prisons admitted to feeling "that their ethics were regularly compromised in their work setting," in particular because "security staff might retaliate if health staff reported patient abuse." A year before the survey was completed, the U.S. Justice Department published a report about the brutalities routinely visited on incarcerated people in New York's main prison, Rikers Island, where "a deep-seated culture of violence" had taken root among the guards. The violence flourished, in part, because of "inadequate reporting by staff . . . including false reporting," the Justice Department concluded. One counselor who did try to report it was Randi Cawley. In 2012, Cawley was sitting at her desk in the mental health assessment unit at Rikers when a group of guards entered with a young man who was handcuffed to a gurney. After dragging him into an examination room with no cameras in it, the guards took turns slugging him in the face with closed fists, a cascade of blows that went on for five minutes. Then the guards got to work on another prisoner, who suffered multiple contusions. In the morning, the walls of the unit were streaked with blood. Yet the official incident report made no mention of abuse. One guard told an administrator the victims had "hit their heads on the cabinets themselves." Cawley decided to report what really happened, naming the specific guards and officers involved. Afterward, she began to field threats: dead flowers were placed on her computer; ominous messages were left on her phone. Eventually, she felt so unsafe inside Rikers that she quit.

The abuse at Rikers was brazen and extreme. But how exceptional was a culture of violence in the mental health wards of America's prisons? Not very, the evidence suggested. In 2015, a report by Human Rights Watch found that excessive force was routinely administered to the 360,000 prisoners in America with mental illnesses. Chemical sprays, stun guns, extended solitary confinement: all of these tools were used with disturbing regularity to incapacitate and punish mentally ill prisoners. In the view of Jamie Fellner, then a senior adviser at Human Rights Watch and the report's author, one reason for this was that mental health professionals rarely intervened to suggest alternatives. "Mental health staff in prisons all too often acquiesce," said Fellner. "There is this culture of 'It's none of our business,' which means that nobody ends up advocating for the patient."

In Fellner's view, the mental health attendants who made these accommodations were not victims. They were enablers, shirking the "duty to protect" and deferring to security in ways that could have fatal consequences.

In one case described in the Human Rights Watch report, a mentally ill prisoner in Dallas died after officers kicked him, choked him, and doused him with pepper spray, even after he had been placed in restraints and had stopped resisting.

Kenneth Appelbaum, a psychiatrist who spent nearly a decade as the mental health director of the Massachusetts Department of Correction, agreed that excessive deference to guards was a problem. But Appelbaum also faulted professional organizations such as the American Psychiatric Association for paying little attention to the ethical challenges facing clinicians who worked in prisons. At the APA's annual meetings, Appelbaum said, "barely 1 percent of the sessions have anything to do with care and treatment in a correctional setting. Prisons are where so many of the sickest people with the most serious psychiatric disorders in our society end up, and as a profession we constantly lament this. Yet our professional organizations are not very engaged in asking how we should care for patients in those settings."

The lack of engagement was all the more striking in light of the scale of the need. In 2014, the Treatment Advocacy Center, a nonprofit organization dedicated to eliminating barriers to care for mental illness, and the National Sheriffs' Association copublished the first national survey of treatment practices in correctional facilities. In forty-four of the fifty states it surveyed, the institution holding the most people with severe mental illnesses was not a hospital. It was a jail or prison. The pattern prevailed in liberal states like California, where the largest mental health institution was the Los Angeles County Jail, and in conservative ones like Indiana, where several state prisons held more mentally ill people than the largest remaining psychiatric hospital. Overall, "the number of mentally ill persons in prisons or jails was 10 times the number remaining in state hospitals," the report found. Among the consequences were overcrowding, physical attacks on corrections officers, an overreliance on solitary confinement, and the deterioration of people in custody who were subjected to neglect and abuse.

"When things go wrong, as they inevitably do, the prison and jail officials are blamed," the Treatment Advocacy Center noted. The real problem rested with a society that failed to fund and maintain "a functioning public mental health treatment system," the report suggested, and that had forgotten its own history. In colonial America, "lunatics" and "mad persons" had often been confined in jails and prisons. But by the middle of the nineteenth century, this practice came to be seen as barbaric and cruel, thanks to reformers like Dorothea Dix, who drew attention to the deplorable conditions that the mentally ill were forced to endure in the jails of Massachusetts and other states. In a report submitted to the Massachusetts legislature in 1843, Dix relayed seeing indigent wards with severe mental health problems "chained, naked, beaten with rods, and lashed into obedience." At the urging of Dix and other reformers, a network of asylums and public psychiatric hospitals soon emerged to provide more suitable treatment. By 1880, the federal census found that "insane persons" accounted for less than 1 percent of America's jail and prison population.

A century later, when a new generation of reformers pushed for state psychiatric hospitals to close, few imagined the horrors that Dorothea Dix had witnessed would recur. One person who did anticipate this was Marc Abramson, a psychiatrist who visited the county jail in San Mateo, California, in the early 1970s, as the deinstitutionalization movement was gaining traction, and noticed the large number of mentally ill people in custody. "There may be a limit to society's tolerance of mentally disordered behavior," warned Abramson in an article published in 1972. "If the entry of persons exhibiting mentally disordered behavior into the mental health system of social control is impeded, community pressure will force them into the criminal justice system of social control." A year after Abramson's article appeared, the California state legislature held hearings to discuss the concerns he had raised. But as the years passed and the number of mentally ill people behind bars continued to grow, the level of public attention faded, suggesting that many communities were happy enough to have jails and prisons take care of a problem that might otherwise have been left to them, just as Abramson had feared. "Perhaps the most alarming aspect of the present situation is that such numbers no longer elicit much professional or public reaction," observed the Treatment Advocacy Center report. "Half a century ago, such reports would have elicited spirited public discussion and proposals for reform; now they elicit a collective public yawn."

In Appelbaum's view, the lack of engagement from professional organizations reflected the fact that the vast majority of elite psychologists had no experience working in jails and prisons and regarded such work as beneath them. "Correctional health care has been seen as the place that clinicians who were not really competent to practice out in the community would end up," he said. "It has had this stigma. If you are working in a correctional setting, there must be something about *you* that you're not up to snuff."

One reason for the stigma was money: the pay in most prisons was paltry, which diminished the prestige and attractiveness of working in them. Another was a body of critical literature produced in the 1960s, when scholars such as Michel Foucault and Erving Goffman published influential studies depicting institutional psychiatry as an instrument of social control. In his 1961 book, *Asylums*, which drew on a year of field research at a psychiatric hospital in Washington, D.C., Goffman wrote scathingly about staff members who "have been reported forcing a patient who wanted a cigarette to say 'pretty please' or jump up for it." Such abuse flourished in what Goffman called "total institutions," regimented places cut off from the outside world where individual autonomy was crushed and where the people in confinement were humiliated and besmirched. The staff who wielded power in such institutions besmirched themselves no less, Goffman's account suggested, a perception reinforced in novels such as Ken Kesey's *One Flew over the Cuckoo's Nest*, in which the villain, Nurse Ratched, exacts revenge on an insubordinate patient, Randle Patrick McMurphy, by arranging for him to be lobotomized (the lobotomy proceeds even though McMurphy is not mentally ill). Such works cast a powerful spell on the popular imagination and made many psychiatrists understandably loath ever to set foot in a "total institution" again. In this context, correctional psychiatry came to be seen not as a laudable form of social work but as dreary and debasing—as dirty work.

"HUMAN MATERIALS"

By the time the *Herald* article on Darren Rainey appeared, Harriet Krzykowski was no longer working at Dade. She wasn't even in Florida anymore. She had gone back to Missouri, where she and Steven had moved so that they could be closer to his ailing mother. When I visited Harriet, she told me that after she left Dade, she tried to erase the experience from her memory. Then she saw the *Herald* story about Rainey's death and the memories came flooding back, a wave of flashbacks that brought on a familiar set of physical symptoms. She couldn't eat. She fell into a depression. She started losing her hair again.

When we met, Harriet was wearing faded jeans, a short-sleeved blouse, and a black wig. A trace of melancholy glimmered in her eyes. Enough time had passed that she was finally ready to discuss her experience, but revisiting the past also made her think about why she hadn't spoken out earlier. "There was one particular night I couldn't sleep because I was crying too hard, thinking, oh my God, all this time has gone by and I didn't say anything, even when I was out of the situation," she said. "I let it continue. These guys are still suffering. They're still there—it's still happening, it's happening all over. Why didn't I do more?"

Harriet was not the only former Dade employee who was haunted by such thoughts. A few weeks after we met, I had lunch in Miami with a behavioral health technician named Lovita Richardson. Originally from Daytona Beach, Richardson told me she was initially excited to work at Dade, convinced that she could make a positive difference in the lives of her patients, whom she saw as victims of societal neglect. "This is an invisible population; people assume you're scum of the earth," she said of the men in the TCU. Richardson knew how easily derogatory assumptions could render people invisible in America. She was raised by her paternal grandparents, who, like many African Americans of their generation, experienced the indignities of the Jim Crow South directly when they worked as domestic servants for a wealthy white family in Miami Beach, where Black people were not allowed after sundown. Growing up, Richardson was the only Black student at the Catholic girls' school she attended. At first, Richardson relished the opportunity to treat the prisoners at Dade with the compassion that she felt they deserved. "I couldn't wait to get to work," she said. One morning, at around 10:30, she walked out of the nurses' station in the TCU, toward the exit, and then stopped. She turned around, circled back, and glanced again at what she thought she'd seen out of the corner of her eye. Across the glassed-in hallway, a group of guards were bludgeoning a prisoner who was shackled to a chair inside his cell, punching and beating him while one of them stood watch. The victim absorbing the onslaught was a tiny man, "maybe 110 pounds soaking wet," Richardson recalled. Richardson watched in stunned silence for several minutes as the battering continued, long enough for the guard on lookout duty to spot her.

Afterward, Richardson wanted to report what she'd seen, but she hesitated when a coworker with more experience warned her that she would merely be endangering herself. In the days that followed, the guards dropped by to tell her that she wouldn't be needed in that unit because they had already taken care of everything. Their tone was polite, but the message was clear. "They let you know, we're running this place—this is *our* house, you're just visiting," she said. Soon thereafter, she started having nightmares and questioning what kind of person she was. When I asked her if she still had these thoughts, tears welled in her eyes. "It makes you feel like you're letting this person down," she said. "They are at risk for their very life, and you know it, but you're not helping them out."

The coworker who advised Richardson also spoke to me, though she did not want to be identified. She understood exactly how Richardson felt, because she had been in the same situation herself. She was the mental health counselor who had called George Mallinckrodt after seeing guards stomp on a prisoner in handcuffs. Afterward, she told me, "I wanted to cry. I wanted to scream." Yet on the form indicating what she had witnessed, "I wrote I didn't see anything." A Latina woman who grew up in East Harlem before relocating to Florida, she, too, felt a responsibility to treat incarcerated people humanely, both because it was her job and because she knew what it was like to experience fear and helplessness in the presence of security officers. (Growing up in Harlem, she said, "I knew that racism was alive and well and that the police, if they got the opportunity, would step on me.") After witnessing the stomping incident, she told me she thought of quitting but, like Harriet, was in no position to leave her job. "If I had not needed my paycheck, I would have walked out," she said, her eyes falling. "There were no other jobs."

The only mental health counselor I met who told me he did not fear for his safety while working at Dade was George Mallinckrodt, who is six feet three, with a broad wingspan and a lanky, athletic build. But Mallinckrodt said the job nearly caused him to have a nervous breakdown. When he was fired, he felt a wave of relief wash over him. The relief eventually gave way to unsettling memories, among them an exchange he'd had with a prisoner who kept flinging his food tray at the window of his cell, as though it were a Frisbee. After failing to persuade him to stop, Mallinckrodt concluded that the man was in the throes of a psychotic episode. He also noted, with surprise, that there were no food stains on the window. Only later, when he heard about prisoners receiving empty meal trays as a form of punishment, did he realize that the man was outraged because the guards were starving him. "I was seeing abuse, but I was labeling it as 'Oh, he's psychologically compromised; he's mentally ill,'" Mallinckrodt said. After leaving Dade, Mallinckrodt published a memoir, *Getting Away with Murder*, that described the cruelty he witnessed. In one passage, a prisoner tells Mallinckrodt about a patient in the TCU who, after receiving an empty food tray, stuck his arm through the flap on his cell door, demanding something to eat. A guard grabbed his arm; then another officer came over and kicked the door flap, smashing the arm again and again. Mallinckrodt talks to the victim, who shows him his bruises, and reports the incident. Nothing is done.

Like George Mallinckrodt, Harriet Krzykowski felt a compulsion to write down her experiences. Halfway through our first conversation, she reached into her purse and handed me a fifty-two-page single-spaced manuscript. She wrote it in feverish outbursts, she said, jotting down details on her arm when she didn't have paper or a laptop nearby. The manuscript wasn't finished and had no title, but she called it her "trauma narrative," a label that matched the diagnosis she'd been given by a therapist she had started seeing, who told her she had post-traumatic stress disorder. In one passage, Harriet recalled seeing a guard repeatedly taunting a prisoner by calling him Tampon, until the man flew into a rage. When Harriet asked the guard about the insult, he said, "He got his ass raped, and now he needs a tampon to stop the bleeding." Harriet later spoke with a nurse, who confirmed that the prisoner had been sexually assaulted.

At our second meeting, Harriet showed me excerpts from a diary that a patient in the TCU had shared with her, scraps of paper that were covered in a looping, childlike script. "I can't escape the feeling that someone is coming for me in this other place in my mind," a passage from the diary read. The prisoner who wrote it was a convicted drug dealer with a range of physical and mental disabilities, Harriet said. At one point, he had been hospitalized for trying to swallow pieces of his wheelchair. He was also a victim of extreme childhood abuse who had lost his wife and two daughters in a car accident, she learned. This won him no sympathy from one of the guards, who called him a "loser" and, one day, tipped him out of his wheelchair.

"This is a real person, with a real life," said Harriet.

Total institutions enabled abusive staff to exercise unchecked power, Harriet had learned. Yet her own experience illuminated another dynamic that could unfold in these institutions: sometimes, bonds could develop between incarcerated people and sympathetic staff. This possibility had not escaped Erving Goffman, who noted at one point in *Asylums* that what transpired in total institutions was "people work," a form of labor that required interacting on a daily basis with "human materials." Even in institutions designed to maximize the social distance between patients and staff, such work could be morally and emotionally discomfiting. "However distant the staff tries to stay from these materials, such materials can become objects of fellow feeling and even affection," Goffman observed. "There is always the danger that an inmate will appear human; if what are felt to be hardships must be inflicted on the inmate, then sympathetic staff will suffer."

Not long after Harriet shared her trauma narrative with me, I met one of her former patients, Harold Hempstead, at the Columbia Correctional Institution in Lake City, Florida. He had been transferred after his sister, Windy, convinced officials at Dade that his life would be in danger if he remained there. We spoke for an hour in a featureless gray room while a sergeant stood watch, during which Hempstead recounted how, after Darren Rainey's murder, several mental health counselors urged him to stop "obsessing" over the crime. One told him he was being "delusional"; another cautioned him to keep any accusations "vague." Hempstead acknowledged the pressure that the mental health counselors in the TCU were under. "Their hands were tied," he said. But too many had internalized the view that the men in the unit deserved rough treatment, because they were criminals or because they were socially marginal. If more counselors had been willing to stand up for the prisoners, he said, "the majority of that stuff wouldn't have happened."

Before I left, Hempstead told me that he had a confession to make. A few weeks before Rainey died, he had informed a guard that there appeared to be dry excrement on a Koran that Rainey owned. The guard seized the Koran and, over Rainey's protests, threw it away. Rainey later confronted Hempstead, calling him an "effing cracker." Hempstead said that he deeply regretted talking to the guard, because losing the Koran triggered the breakdown that made Rainey a target of prison staff.

The only therapist who had helped him work through the guilt he felt about Rainey's death was Harriet, whom he affectionately called Ms. K. "She would actually listen," he said. "She attempted to enroll me in some classes dealing with trauma." He paused. "I really didn't like to see her go."

In January 2016, the Miami-Dade medical examiner delivered a copy of Rainey's autopsy report to state prosecutors. The report was not made public, but its contents were leaked to the media. It concluded that the guards at Dade had "no intent" to harm Rainey and that his death was "accidental." No criminal investigation was recommended, notwithstanding the fact that Rainey's death was not an isolated incident. At least eight other people in the TCU endured abuse in a scalding shower. Among them was Daniel Geiger, who was subsequently transferred to the Lake Correctional Institution, near Clermont, Florida. When I spoke with his mother, Debra, who lives in North Carolina, she told me that she had not seen her son in several years, because prison officials had denied him visitation rights, claiming that he was dangerously unstable. She said that she had last talked to Daniel in 2012, on the phone, shortly after he was transferred out of Dade. It was a brief conversation, during which he spoke with a slur and could not pronounce simple words. He told her that his weight had dropped from 178 pounds to 105. Although she was alarmed by this news, she did not suspect that he had been abused, only that "something was being hidden" from her.

I told Debra Geiger that, according to several sources, her son had been forced multiple times into the same shower where Rainey had died. He was also among the prisoners who had been denied meals. "I'm heartbroken," she said, her voice cracking. When we spoke again a few days later, she told me that she had called the Lake Correctional Institution and learned that her son was being given two psychiatric medications to which he was allergic. Later, she sent the facility a note from her son's former psychiatrist, Dr. James Larson, confirming the allergies. She received a two-line response, saying that the information would be forwarded to the medical staff, but she was not told if the treatment had been discontinued. "It's hard for me to digest all of this," she said, comparing the treatment of mentally ill incarcerated people to the torture of detainees at Guantánamo.

In late February 2016, Geiger was finally permitted to visit her son. She described him as being "at death's door," a gaunt figure with sunken eyes who mistook her for his wife and growled at the guards when they called his name. His arms were skeletal—"no wider than my wrists," she said—and there were deep red marks on his neck. When she asked him what had happened at Dade, he peered up at the ceiling, pressed his face against the glass partition separating them, and said, "I don't want to talk about it." She said that before he was sent to prison, when her son was taking the right combination of medications, he was relatively stable—a point that she impressed on the warden before leaving. "I told him that I kept him more or less normal for thirty-three years, and you all have destroyed him in seven," she said.

In September 2014, Disability Rights Florida, an advocacy organization, filed a lawsuit charging the Florida Department of Corrections with subjecting mentally ill people at Dade to "abuse and discrimination on a systematic and regular basis." According to the terms of a settlement reached one year later, the Department of Corrections agreed to make several changes, including the installation of a new camera system at Dade, better training of guards, and the hiring of an assistant warden of mental health.

Not long after the settlement was announced, I drove to Dade to meet the new assistant warden, Glenn Morris. The prison sits on the outskirts of the Everglades, on a two-lane road flanked by squash and tomato fields and by signs advertising alligator farms and airboat rides. After exchanging greetings, I followed Morris into a control room, through a metal detector, past a heavy steel door that opened onto a quarter-mile cement path that led to a cream-colored building with a sign above the entrance: TRANSITIONAL CARE UNIT. The path wasn't shaded and, although it was early, the heat was stifling. Much to my relief, the air-conditioning inside the TCU was functioning (Harriet told me that when she worked there, it was often broken). The walls were a dull gray, but the interior looked clean: an orderly in faded scrubs was sweeping the concrete floor. We walked down a nurses' hall and stopped at the entrance portal to the west wing, a cavernous chamber furnished with metal tables and lined with single-person cells, each of which had a small rectangular window. There were no prisoners inside; they were out in the rec yard, Morris told me. He pointed to several TV sets that had been installed recently and to the mural on the far wall, a cheerful ocean scene that had been painted by a prisoner.

"When I first got here, the mentality was, 'This is confinement,'" Morris told me. "I had to change that." The Department of Corrections had adopted an "open cell" policy at Dade, he said, so that lower-risk prisoners could move around more freely.

Morris introduced me to the unit's major, a large man with a broad smile, and to several corrections officers. All of them were recent hires. The new faces were part of the overhaul, he said. When we crossed over to the east side of the TCU, which housed people who were deemed more stable, a meeting was ending, and Morris introduced me to a more experienced staff member, Dr. Cristina Perez. "She does a very good job," he said. Before visiting Dade, I had actually called Dr. Perez to ask her about some of the problems Morris had been hired to rectify. She refused to answer any questions. Dressed casually in sneakers and sweatpants, Dr. Perez was more cordial in person. She extended her hand after Morris's introduction. "Nice to meet you . . . oh," she stammered, evidently recalling that we had spoken before. She smiled uneasily and walked away.

Afterward, in an administrative office, I asked Morris if the desire to appease security might compromise how well mental health counselors did their jobs. "Dr. Perez, I'm sure, tells her staff to report things to her," he said. "And I'm very confident if she found out something, she would report it to us." I mentioned that I had heard otherwise. Morris rolled his eyes, telling me he assumed that my understanding came from a "disgruntled" ex-employee nurturing a grudge—meaning George Mallinckrodt. I told him that other former employees had voiced similar misgivings. "Obviously, that was way before my time," he said.

Morris came across as sincere and well intentioned, but his assurance that prisoners were getting their "basic needs" met was disputed by other sources. Prisoners still came to the inpatient unit of the TCU and languished after being placed in what effectively amounted to solitary confinement, I was told. Many patients received no real treatment at all. In one case that was described to me, a man afflicted with paranoia had been degenerating steadily for more than a year. Though he was not disruptive, he'd spent prolonged periods in lockdown because he had stopped taking his medication. Nobody had encouraged him to try a different medication or engaged him in activities that might have lessened his feelings of distrust. As a result, one source said, the patient was "undergoing a quiet decompensation where he just gets sicker and sicker."

The mental health staff at Dade continued to defer to security, acquiescing when their clients were given disciplinary reports for conduct clearly related to their illnesses rather than intervening to suggest treatment. A man with diagnosed impulse-control problems had his privileges taken away after an anger outburst, for example. Following a cursory discussion, mental health officials checked a box indicating that his issues had played no role.

According to Bob Greifinger, a professor at the John Jay College of Criminal Justice who studies mental health conditions in prisons, such routine neglect is no less pernicious than instances of flagrant abuse, not least because it is rarely perceived as such. "Most of the coercion that happens goes relatively unrecognized," Greifinger told me. "There are very few people who can step back and say, 'Hey, wait a minute. They are trying to interfere with my taking care of my patients.'" One observer who had sat in on a staff meeting at Dade said that the counselors and psychiatrists seated around the table seemed more oblivious to the mental health needs of the prisoners than the corrections officers in the room.

In the fall of 2015, the Civil Rights Division of the Justice Department launched an investigation to determine whether the death of Darren Rainey was part of a broader pattern of abuse. It was not the first time the Florida prison system had been subjected to federal scrutiny. In 2006, James V. "Jimmy" Crosby, who had been appointed DOC secretary by the then governor, Jeb Bush, was sentenced to eight years in prison for accepting kickbacks from private vendors. Under Crosby, corruption was not just tolerated but rewarded, Ron McAndrew, a former warden, told me, with a blind eye turned to "goon squads" who regularly beat up prisoners for sport. (Crosby himself was the warden of a prison where, in 1999, a prisoner named Frank Valdes was brutally beaten to death.) The culture changed under James McDonough, a West Point graduate and former Vietnam platoon leader who succeeded Crosby and dismissed dozens of wardens and senior officials he suspected of corruption. But McDonough's tenure lasted little more than a year. Meanwhile, tougher sentencing laws filled Florida's prisons beyond capacity, even as budget problems led to reductions in staffing, forcing guards at many prisons to work longer shifts that raised stress levels and, in turn, the likelihood of abuse.

Julie Jones, the DOC secretary hired in the wake of Darren Rainey's murder, vowed to change the culture of Florida's prisons again. A few months after taking office, Jones issued a "statement on retaliation" in which she vowed that no employee "who comes forward with an issue of concern would face retaliation." On the wall of the main administrative building at Dade, I spotted a framed copy of this statement. A few months earlier, however, Jones had circulated a memo requiring all DOC inspectors to sign confidentiality agreements about the investigations they conducted. There was no sign of this memo on the wall at Dade.

After relocating to Missouri, Harriet Krzykowski briefly contemplated working at another prison. She ended up finding a job at an agency for at-risk youth, counseling children who had been exposed to traumatic abuse and violence. Not infrequently, the experiences of the children reminded her of her own traumatic upbringing. They also reminded her of the stories she'd heard from patients at Dade, many of whom had been subjected to harrowing violence from an early age. Violence begat violence, she came to believe, producing criminals who victimized others because, in many cases, they had been victimized themselves. She was not alone in being struck by this. In 2012, researchers in Boston began conducting in-depth interviews with 122 incarcerated people in the Massachusetts prison system. Half of the subjects reported that their parents had beaten them. Many had been sexually assaulted and witnessed shootings in the chaotic, unsafe neighborhoods in which they grew up. "Violent offenders, more often than not, are victims long before they commit their first crime," observed the sociologist Bruce Western, who helped lead the study. The pervasiveness of past victimization underscored why all prisoners deserved not just punishment but "mercy and compassion," Western argued, something few encountered behind bars. Three-fourths of the subjects in the study reported seeing an assault by prisoners or staff.

Harriet found her new job rewarding, but she remained haunted by her experience. One morning, I met her at a nature reserve on the banks of the Missouri River that she liked to visit on weekends. We strolled along a winding, leaf-strewn trail for half an hour before stopping at a pagoda by the river, a toffee-colored ribbon along which a couple of rowboats drifted. Harriet sat on a bench that overlooked the gently churning water. She propped her feet up on a ledge. Before working at Dade, she told me, she'd had a clear view of morality. "I always felt there was a right and there was a wrong and that was all there is to it," she said. "People who are in the position to do the right thing are always going to do the right thing because that's just who they are; that's why they're drawn to that side." Now her view was muddled and blurred, washed in the gray light of her experience. Was she a victim of the system, or an instrument of it? Whose side had she been on? Sometimes, she reminded herself that she'd had no good choices; she was "a nobody" at Dade, "the lowest person on the rung" in a place where the slightest sign of insubordination was dangerous. Other times, she would think this was just a rationalization and feel a quiver of guilt or shame, as on the evenings when she drew a bath for her kids and checked the temperature of the water, causing her to think of Rainey.

Perhaps because it afforded more privacy, Harriet seemed to prefer meeting outdoors, in parks and conservation areas where no one could eavesdrop on our conversations or see her crying, which happened more than once. But one afternoon, she suggested we meet indoors, at a place called the Glore Psychiatric Museum in St. Joseph. The museum occupies a drab brick building that looks like a hospital. After passing through a small gift shop on the ground floor, we climbed a flight of stairs that led to a series of rooms filled with arcane medical devices—a fever cabinet, a lobotomy table—that offered an unsettling commentary on how people with mental disabilities were treated in the past. At one point, Harriet stood before a life-size replica of a cell where patients were once held at the Hospital of Salpêtrière in Paris. Michel Foucault wrote about the hospital in his 1961 book, *Madness and Civilization*, during the era he called "the great confinement," which began in the mid-seventeenth century. Harriet peered into the cell, a dingy chamber littered with straw. She stopped to read the label tacked on the wall:

> *At the hospital of Salpêtrière, the insane were kept in narrow filthy cells . . . When frostbite resulted, as it often did, no medical help was available. Food was a ration of bread once a day, sometimes supplemented by thin gruel. The greatest indignity was the chains.*

Afterward, in the gazebo outside, Harriet reflected on how little had changed. "We don't learn very fast," she said.

As I subsequently found out, the Glore Psychiatric Museum is situated in the medical wing of a former state psychiatric hospital. Curious to see the main compound, I went back a few days later. Following a narrow walkway shaded by pines, I arrived at a more secluded area, farther away from the main road, where the hospital's residential quarters once stood. The path ended at a security barrier, a chain-link fence topped with razor wire that encircled what the shuttered hospital once known as the State Lunatic Asylum No. 2 has become: a prison.

2

The Other Prisoners

Prodded by media coverage, the State of Florida eventually re-opened an investigation into Darren Rainey's death to determine whether the security officers who had subjected him to the "shower treatment" should be prosecuted. On March 17, 2017, Katherine Fernandez Rundle, the state attorney for Miami-Dade County, announced that the investigation had concluded and that no criminal charges would be filed against any of the guards involved. According to a report issued by Rundle's office, formally titled "In Custody Death Investigation Close-Out Memo," Rainey's death was caused by schizophrenia, heart disease, and "confinement inside a shower room," not by mistreatment or criminal negligence. "The evidence does not show that Rainey's well-being was grossly disregarded by the correctional staff," it stated.

To reach this conclusion, the Rundle report dutifully ignored the voluminous evidence that gross disregard was pervasive at Dade. Nowhere did the report discuss the climate of abuse and intimidation that Harriet Krzykowski and other mental health counselors witnessed. Nowhere did it mention that prisoners were beaten, tortured, and starved. State investigators did collect testimony from some critical sources, including Harold Hempstead. But Hempstead was portrayed as an unreliable witness whose statements contained inconsistencies and whose allegations "tainted" the views of other prisoners (several of whom corroborated his account). Far less skepticism was applied to the testimony of the guards, who, investigators concluded, took Rainey to the shower out of concern for his hygiene, "allowing him to wash himself" after they had spotted feces smeared on his body and on the walls of his cell.

Investigators were equally unquestioning of Dr. Emma Lew, a medical examiner who told them that Rainey "did not sustain any obvious external injuries and, particularly, that there were *no thermal injuries (burns)* of any kind on his body." This conclusion was at odds with an interim report that the Miami-Dade medical examiner filed immediately after Rainey's death, in which an autopsy technician had observed, "Visible trauma was noticed throughout the decedent's body." (The interim report was not mentioned in the "Close-Out Memo.") Dr. Lew's conclusion was also at odds with a series of autopsy photos of Rainey's body that were soon leaked to the press. One of the photos showed Rainey's chest, a grisly patchwork of raw white tissue and exposed blood vessels. Another showed his back, which was flayed from the nape of the neck down to the shoulder blades. Rainey had visible wounds on his arms, legs, face, and stomach. Two forensic pathologists who examined the pictures told the *Miami Herald* they believed his death was a homicide and that the injuries he sustained were burns.

The Rundle report was a whitewash, illustrating the culture of impunity that prevailed in Florida when it came to crimes committed by guards. "We are appalled that the state attorney did not look deeper into this case and see the criminality of the people who were involved," Milton Grimes, an attorney for the Rainey family, told the *Herald*. Harriet Krzykowski was equally appalled. Yet as dismayed as she was, Harriet did not seem to blame the guards at Dade for the cruelty and degradation she witnessed. Many of the officers she met were decent people who treated the prisoners with respect, she told me. And the ones who were abusive, she maintained, acted in ways that were to be expected in a society that had resumed warehousing mentally ill people in prisons, as if they were beyond hope.

The more brutal guards were doing society's dirty work, in other words, which, as it happens, is exactly what Everett Hughes suggested in his 1962 essay. "From time to time, we get wind of cruelty practiced upon prisoners in penitentiaries or jails," Hughes observed at one point in the essay. The impulse to blame such conduct on the custodians in charge was natural. But the people who ran prisons were "our agents," Hughes argued, dispensing punishment to an "out-group"—convicts—that much of the public felt "deserve something, because of some dissociation of them from the in-group of good people." If the punishment in question was "worse than what we like to think about, it is a little bit too bad. It is a point on which we are ambivalent."

Hughes proceeded to consider the matter from the perspective of a "minor prison guard" who had engaged in some "questionable practices" and boastfully justified his conduct afterward, sneering at the hypocrisy of the "higher-ups" and "good people" who held him in contempt. The guard had good reason to see them as hypocrites, Hughes averred. "He knows quite well that the wishes of his employers, the public, are by no means unmixed. They are quite as likely to put upon him for being too nice as for being too harsh. And if, as sometimes happens, he is a man disposed to cruelty, there may be some justice in his feeling that he is only doing what others would like to do, if they but dared; and what they would do, if they were in his place."

"KEEPERS"

That prison guards received mixed messages from society was not news to Bill Curtis. In 2004, Curtis bought a black leather belt at Walmart, put a spit shine on a pair of 1969 military-issue boots—the pair he brought back with him from Vietnam, where he served for two years in the Twentieth Engineer Brigade (the same unit as a future vice president named Albert Gore)—and began working as a guard at the Charlotte Correctional Institution in Punta Gorda, Florida. Within a week, Curtis traded in his heavy army boots for a pair of lightweight nylon ones—being quick on your feet was essential in his new job, he'd discovered—but the memory of serving in the military was rarely far from his mind in the years to come. Working at a prison, he came to realize, was not unlike stepping onto the battlefield in a combat zone.

Originally from southern Illinois, Curtis had moved to Florida with his wife in 1989. For the next fifteen years, he worked in retail, at a furniture store, making decent money until a dispute with his boss led him to quit. A short while later, he spotted a newspaper ad soliciting applications for corrections officers and decided to respond. Most of the would-be officers in Curtis's training cohort were young men in their twenties or thirties. Curtis was older, with tousled brown hair that had started to go gray, but he was lean and athletic, an amateur boxer with a spry manner who sometimes sparred with younger guys at the gym and held his own.

It did not take long for Curtis to have his sparring skills tested at his new job, or to internalize the "us against them" mentality that pervaded the profession he'd entered: corrections officers on one side, prisoners on the other. The officers had "the power and the keys," the prisoners "the numbers and time," Curtis told me over lunch one day. Neither had much sympathy for the other. When it came to some incarcerated people, little sympathy was warranted, Curtis came to feel. Part of his initiation into life as a guard was learning that if he turned his back to the wrong person at the wrong moment, he could wind up in an ambulance or a coffin. "I've picked up a guy with a slashed throat," he said. "I've fought for my own life a few times." Once, a particularly menacing prisoner who was running contraband into CCI and threatening the staff told him that he needed to "take a long vacation" after Curtis stood up to him. The prisoner ended up taking another officer hostage and pressing a nine-inch knife to his neck. (The officer survived after the assailant was persuaded to hand over the knife and release his captive and then, to Curtis's relief, was transferred to another facility.) To work as a corrections officer was to live in a state of "constant apprehension," Curtis told me, particularly at a place like CCI, which at the time was a "close management" lockdown facility, housing some of the most violent criminals in the state.

Like most of his peers, Curtis came to believe that the system was stacked against corrections officers, not prisoners. "They've got so many rules protecting inmates, and you have to give them so many rights," he groused. Yet he did not deny that corrections officers sometimes bent the rules to their advantage. At our lunch, Curtis pulled out a journal that he'd kept while working at CCI and handed it to me. In 2005, when Curtis was a year into the job, the journal contained an entry in which he described feeling "somewhat astounded" after reading the regulations on the use of chemical agents, which were sprayed at prisoners seemingly all the time. "The regulations state that chemical agents are not to be used as punishment but merely as a final resort in the control of disorderly conduct," Curtis wrote. "The uses of agents I have observed have in large [part] been premeditated and planned punishments." In another entry Curtis noted, "Abuses of power and undesirable behavior are more the norm than the exception among some correctional officers . . . Baiting, teasing, threatening, tricking, and at times physical abuse are common." Later in the journal, he described the case of an officer who had been "caught spraying a mixture of bleach water into the face of an inmate who was known to be allergic or reactive to chemicals." It was not the first time the officer in question had amused himself in this manner.

"I've seen a lot of bad stuff in life—ugly things," Curtis told me over lunch. "But I've never seen real cruel things like what you see in a prison— real cruelty, just intentional cruelty. It's like husbands who beat their wives. That's how some of these guys beat their inmates." Curtis did not hide his disgust for such officers, whom he called "serial bullies." He also made it clear that blowing the whistle on them would have been unthinkable. If a corrections officer ratted out a peer, "you'll be out of work, they'll find a way to fire you, slash your tires, put the blackball on you, nobody will talk to you, or they'll leave you hanging somewhere and tell an inmate to get you," he said.

Curtis wanted to live, and he wanted to keep his job until the age of sixty-two, when he could retire. But there was something else that kept him from getting too vocal or self-righteous about the "serial bullies" who entered his profession. The citizens of Florida got what they paid for, he became convinced, filling their prisons to capacity, running them on the cheap, and then expressing shock and disapproval when presented with the unpalatable consequences. "The real problem here is that our rank and file citizens do not want to pay what it takes to care for their prisoners," Curtis observed in his journal. "We talk a good game about how humane and decent our society is but when it comes down to it the bucks are not there."

Over lunch, Curtis elaborated, pointing out that in 2016 the starting salary for a corrections officer in Florida was twenty-eight thousand dollars. Despite the efforts of Teamsters Local 2011, which represented thousands of Florida COs,* the last raise they'd received had been in 2005. Meanwhile, line staff had been pared to the bone in one facility after

---

* Not all COs belonged to the union, because Florida was a right-to-work state.

another, thanks in part to the Florida governor Rick Scott, who, in 2010, ran for office promising to cut the DOC's budget by 40 percent. After his election, Scott set about privatizing services, slashing jobs, and transitioning COs from eight- to twelve-hour shifts, changes that vastly increased staff turnover and led to a sharp rise in use-of-force incidents. Curtis admitted he hadn't always used force judiciously himself. One time, he "got physical" with a prisoner, body-slamming him hard into the ground, a slab of bare concrete that could easily have fractured his skull. Curtis lowered his eyes as he told me the story. "It was totally illegal," he said. "I shouldn't have done it." But he relayed the story for a reason, in order to drive home the point that even decent officers did bad things in a system that skimped on training, salaries, staffing, rehabilitative programs. It didn't help that so many of the people inundating Florida's prisons belonged in psychiatric hospitals, Curtis added. At one point in his journal, he made note of the number of prisoners at CCI who were "not in their proper mind," thanks to "the drastic reduction in the state run mental health facilities." Curtis had received no training on how to interact with mentally ill people.

A boxer, a military veteran: in these respects, Curtis fit the popular image of the prison guard as a hard-boiled type, a muscle-bound enforcer given to dispensing bruising punishment. Yet as his journal showed, he was not incapable of registering shock at punishment that was unduly harsh, even when the person meting it out was a fellow officer. One of Curtis's hobbies was stargazing. Another was reading (he particularly loved the crime novels of the Florida native Carl Hiaasen). He'd even taken a stab at writing a novel himself, a copy of which he later shared with me. The story was set in a prison teetering on the brink of collapse from budget cuts: the cells overheating because the air-conditioning was busted; the meals larded with cheap protein substitutes in place of beef and chicken; the switch panel constantly shorting out. The main character in the book, Sergeant Bernie Petrovsky, is a veteran of the Iraq War with a fondness for chewing tobacco. Petrovsky hasn't had a raise in six years or a moment to relax because of chronic staff shortages. In one scene, he likens his experience in Iraq to his experience behind the walls. "In the desert war we were being blown up in our Humvees until they decided to spend the money to armor the vehicles better . . . Here the management just keeps cutting the budget, which means less personnel and equipment to maintain and perform to an increasingly higher level of expectations."

As Curtis acknowledged, the novel was a thinly disguised account of his experience. Cheaply run prisons were a lot like cheaply run wars, he had concluded, lowering morale, heightening tension, and leading front-line officers to rely increasingly on the one tool at their disposal: brute force. "If you don't have a lot of money, the only way you can control the unit in prison is through brutality, threats, and fear," Curtis told me. "And in order to do that, every once in a while, you gotta kick a little butt. You gotta get brutal, and your officers learn to do that, just like I learned to do that."

Nobody told Curtis and his fellow guards to get brutal. But no one really needed to tell them this. It was enough to pay them modest salaries to enforce order in overcrowded, understaffed prisons that were neither equipped nor expected to do much else. When Curtis was in Vietnam, most of the officials in charge of America's criminal justice system still subscribed to the rehabilitative ideal—the notion that in addition to maintaining public safety, criminal sanctions should be designed to improve the life chances of convicts, who deserved an opportunity to become productive members of society after completing their sentences. For much of the twentieth century, this was the dominant view among, policy makers and corrections officials, shaping everything from sentencing laws to probation systems to the therapeutic programs offered to prisoners. But by the time Curtis had moved to Florida in the late 1980s, the rehabilitative ideal had given way to a more punitive approach, emphasizing the imperative to punish and incapacitate both violent criminals and nonviolent offenders.

As the rehabilitative ideal crumbled, America's prisons experienced a spectacular boom, thanks to draconian policies—mandatory minimum sentences, three strikes laws—that, for several decades, were popular. The elected officials who enacted these policies spanned the political spectrum, from Ronald Reagan, who vastly expanded the scope and severity of the drug war, to Bill Clinton, whose 1994 crime bill gave states incentives to put even more people behind bars. These laws were not foisted on the public against the wishes of ordinary citizens. They reflected and embodied popular sentiment. In the decades when the rehabilitative ideal held sway, convicts had often been depicted as disadvantaged individuals who'd been unjustly deprived of education and opportunity. Now they were labeled "thugs" and "superpredators," dehumanizing, racially coded terms that implied they were beyond redemption and deserved to suffer. If politicians during this era feared anything, it was not being overly punitive but being perceived as "soft on crime." The shift in rhetoric sent the custodians of jails and prisons a clear message that society expected them to treat prisoners harshly, even if "good people" preferred to be spared the unsavory details and to avoid being reminded that a vastly disproportionate share of those getting locked up were Black men, who were incarcerated at a higher rate in America than in apartheid South Africa. By the time some lawmakers began to question whether this was sensible or just, more than two million Americans languished behind bars.

Another message the custodians of jails and prisons had long been sent was that their jobs were lowly and disreputable. The wages of "turnkeys" and "keepers," as prison guards in America were once popularly known, had always been paltry, their hours long, their penchant for cruelty notorious, a perception as old as the job itself. In 1823, an account of a prison in New York portrayed the guards working there as "small-minded, intoxicated with their power, vulgar." A century later, during the Progressive Era, penal reformers drew up ambitious plans to refashion prisons into more humane institutions. The stark, regimented penitentiaries of the nineteenth century—places like the Auburn State Prison in New York, where convicts were forced to eat their meals in silence and routinely flogged—would give way to enlightened sanctuaries replete with baseball diamonds, workshops, and theaters. But as the historian David Rothman has shown, few prisons were provided with the resources to fulfill this vision, which is one reason that most correctional institutions remained "places of pervasive brutality," staffed by undertrained, overworked guards plying a trade that was "a last resort for the unskilled and uneducated." In his 1922 book, *Wall Shadows*, the criminologist Frank Tannenbaum identified "the exercise of authority and the resulting enjoyment of brutality" as "the keynote to an understanding of the psychology of the keeper." Drawing partly on his own experience serving time on Blackwell's Island, where he was incarcerated for "unlawful assembly" after participating in protests for the unemployed, Tannenbaum attributed this psychology not to character flaws but to structural conditions, in particular the desire that guards felt to affirm that they were different from—and superior to—the people under their watch, with whom they shared the same harsh, oppressive environment. "For his own clear conscience's sake the keeper must, and does instinctively, make a sharp distinction between himself and the man whom he guards," Tannenbaum wrote, "and the gap is filled by contempt." In his classic study *The Society of Captives*, published in 1958, Gresham M. Sykes contested the view that most guards were "brutal tyrants," arguing that to maintain order, many learned to negotiate and compromise with prisoners. But Sykes did not make the life of the typical "keeper" sound any less dreary. "The job of the guard is often depressing, dangerous, and possesses relatively low prestige," he wrote.

Sykes's study appeared in the Eisenhower era, when the overwhelming majority of "keepers" in America, like police officers and other law enforcement agents, were white. In the early 1970s, this began to change, thanks in part to the political ferment of the 1960s and to events such as the 1971 rebellion at the Attica Correctional Facility in New York, which drew attention to the racism that pervaded the nation's prisons and led reformers to call for hiring more people of color to help quell the unrest. One of the demands of the prisoners at Attica was "a program for the recruitment and employment of a significant number of black and Spanish-speaking officers." In 1973, the National Advisory Commission on Criminal Justice Standards and Goals urged commissioners to recruit minority officers with more sympathetic attitudes to prisoners. "Black inmates want black staff with whom they can identify," the commission argued. To test the hypothesis that hiring Black staff would improve relations with incarcerated people, the sociologists James Jacobs and Lawrence Kraft surveyed guards at two maximum-security prisons in Illinois that had begun to diversify. The Black guards in the survey tended to believe that "fewer prisoners belong in prison" and to have more liberal political views, they found. But they were also more likely to agree that punishment was the primary purpose of confinement and to be "more active disciplinarians" than their white peers. One potential reason for this was that administrators screened out Black candidates who were sympathetic to prisoners, Jacobs and Kraft speculated. Another was that, like their counterparts in the police, Blacks who were hired felt added pressure to prove that they belonged by suppressing their sympathies. Whatever the case, Jacobs and Kraft questioned the assumption that hiring minorities would lead incarcerated people to be treated with greater respect, concluding that the attitudes and behavior of guards were "built into the organization of the maximum-security prison." Notably, although they adapted to the strictures of the job, the Black guards in their study were twice as likely to say that they were "embarrassed" to tell people what they did for a living. Also striking was one finding that transcended racial lines. "A majority of guards of both races would not like to see their sons follow their occupation," Jacobs and Kraft found.

Between 1960 and 2015, the percentage of Black prison guards increased more than fourfold, paralleling the growth in the number of Americans—in particular African Americans—behind bars. As the prison population ballooned during the age of mass incarceration, Blacks were increasingly given the "opportunity" to run the penal institutions where more and more people of color were caged. In many urban areas, the makeup of the guards came to mirror the makeup of the men and women in custody—places like New York City, where, by 2017, Blacks and Latinos made up nearly 90 percent of the correctional staff. The ranks of female officers also surged. These were "good jobs," some economists argued, and it is true that the growth of America's prisons lent prison guards—newly christened "corrections officers"—a degree of newfound legitimacy. In states like New York and California, COs earned decent salaries and joined unions that came to wield substantial political clout. In some rural areas, working at a prison soon became the best employment prospect around, offering benefits unavailable at fast-food restaurants or in the mills and factories that had long ago left town.

But the jobs in question tended to be reserved for people with limited options who lived in struggling backwaters. A few days after meeting Bill Curtis, I drove to one such place—Florida City, the last stop on Florida's Turnpike before Key West and, along with neighboring Homestead, the area that furnished Dade with most of its rank-and-file guards. Other than to fill up on gas before racing off to the Keys, more affluent Floridians rarely frequented places like Florida City. When I visited one Friday morning, the town looked sun beaten and deserted. The dingy storefronts and random businesses lining the main commercial thoroughfare were empty or shuttered. A mile or so down the road, I pulled in to a plaza that was unusually busy. It was situated across the street from a church with a tattered sign out front—PREPARE TO MEET THY GOD—and was home to at least one thriving local establishment, a yellow building with paint-chipped walls that turned out to be a welfare agency. Near the entrance was a poster of a Latina girl in pigtails holding the halves of two oranges to her ears and smiling, beneath the words WIC: GOOD NUTRITION FOR WOMEN, INFANTS & CHILDREN.* Inside, several women with toddlers were waiting in line. None of them were smiling. In Florida City, 40 percent of families fell below the poverty line. Many of the entry-level guards at Dade were Black and Latina women, Bill Curtis told me, in part because so many young men of color from Homestead and Florida City had police records (which disqualified them from working for the DOC). "They need jobs, they've got kids," Curtis said of the female guards at Dade, "and it's the best job they can get."

The area surrounding Dade was an example of what the sociologist John Eason has called a "rural ghetto"—small towns in depressed rural areas that were home to many of the prisons built in America since the 1970s, when the number of correctional facilities tripled. Until this time, rural areas tended to oppose allowing prisons to be constructed on local land for the same reason that wealthy suburbs did: to avoid association with institutions that were seen as disreputable and potentially dangerous. But as factories closed and family farms went bankrupt, the civic leaders in many rural areas began lobbying to have prisons built in their counties. Whether any lasting economic benefits resulted from this strategy is unclear (one study concluded that, to the contrary, prisons *impeded* growth in the areas where they were located). What is clear is

---

* WIC is a supplemental nutrition program for low-income women.

that luring these "stigmatized institutions" to town further cemented the lowly status of the communities in question, places of concentrated disadvantage where poverty and racial segregation were deeply entrenched. "Stigmatized places are more likely to 'demand' stigmatized institutions, particularly if the stigma of the community is equal to or greater than the stigma associated with the institution in question," Eason observed. "Rural towns most likely to receive a prison suffer the quadruple stigma of rurality, race, region, and poverty."*

After leaving the plaza in Florida City, I drove around town, stopping to talk to a man named Jimmy who was sitting in a plastic lawn chair by the side of the road, next to an empty shopping cart. An African American man with a Phillies cap on, Jimmy was selling fruits—mangoes and bags of lychees that he'd arranged on a rickety table. How were things? I asked. "Hard," he said. Jimmy had lived in Florida City for twenty-five years. He appreciated the sunshine and warm weather but was less fond of the business climate, which he told me was moribund (I was his only customer). When he was young, Jimmy said, he picked okra and tomatoes in the fields. The migrant workers now were "mostly Mexicans and Haitians 'cause it's hard work," he told me. A little girl in dreadlocks soon came out to join us. She was Jimmy's granddaughter and lived in the dilapidated housing complex behind his makeshift fruit stand. Next to the housing complex was an abandoned lot strewn with broken glass and garbage.

I bought some mangoes from Jimmy and drove on, past a Laundromat and a dollar store, past a strip of row houses, past some squash and bean fields, which looked parched and wilted in the scorching sun. I turned left at a fruit stand called Robert Is Here and, after a few more miles, saw the soaring gray floodlights surrounding Dade CI. The prison squatted in the distance, a cluster of dun-colored buildings set behind a maze of fences topped with concertina wire. In front of the outermost fence, I spotted a sign:

---

* Although the term "rural" is often associated with white regions such as Appalachia, Eason found that the southern towns where prisons are concentrated have a *higher* proportion of Black and Hispanic residents than the areas surrounding them.

It was the only notice for jobs that I came across in Florida City.

COPING MECHANISMS

In her 1973 book, *Kind and Usual Punishment*, the muckraking journalist Jessica Mitford wrote, "For after all, if we were to ask a small boy, 'What do you want to be when you grow up?' and he were to answer, 'A prison guard,' should we not find that a trifle worrying—cause, perhaps, to take him off to a child guidance clinic for observation and therapy?"

Mitford's question betrayed a common preconception, which is that anyone who yearned to become a prison guard was a bit morally suspect. It also betrayed a dubious assumption, which is that becoming a guard was something people aspired to do, as opposed to something they stumbled into or got stuck doing for lack of better options. In 2010, a team of scholars conducted a survey that compared why police officers and prison guards chose their careers. While the police officers emphasized "service as a primary factor," the corrections officers "placed greater importance on financial motivators," including "a lack of other job alternatives." None of the guards interviewed by the sociologist Dana Britton in her book *At Work in the Iron Cage* "grew up dreaming of working in prisons or even planning to do so." Most entered the profession after a period of "occupational drift," with few positive aspirations and little idea of what they were getting into.

It's no wonder, considering what the occupational health literature suggested lay in store for prison guards: alarming rates of hypertension, divorce, depression, substance abuse, suicide. One study in New Jersey found that the average life expectancy of a corrections officer was fifty-eight. Another, drawing on data from twenty-one states, found that the suicide risk among corrections workers was 39 percent higher than for the rest of the working-age population.

I'd met corrections officers whose stories bore out these dispiriting facts. There was a CO from New England I'll call Johnny Nevins (he did not want his real name used) who went home one day, drank an entire bottle of whiskey, posted a video on Facebook saying goodbye to his family, and then pointed a loaded gun at his head. The cartridge jammed when he pulled the trigger, and Nevins had since launched a support group to help troubled COs avoid going down a similar path. He'd also begun seeing a therapist, who diagnosed him with acute PTSD caused by repeated exposure to extreme violence.

There was Tom Beneze, a guard from Cañon City, Colorado, a town of sixteen thousand that was home to a Walmart, some fast-food chains, and more than a dozen jails and prisons. When I met Beneze one morning on the back porch of his modest ranch house, he reached into his pocket and pulled out a box of pills that he kept with him at all times to ward off the anxiety that periodically rippled over him. "If I start buzzing real bad, I'll take one of those," he said, pointing to the small white tablets in one of the compartments, which helped ward off panic attacks. Next to the white tablets were some blue capsules he took for depression. A third slot was filled with painkillers for his various injuries, including nerve damage sustained when a prisoner wielding a shank gashed his leg. Beneze, who had a thinning patch of reddish-brown hair and an anxious, watchful air, told me he'd seen "more hand-to-hand combat" than his son, a navy veteran who'd done two tours of duty in Iraq. He never sat in a restaurant with his back to the door, a habit ingrained in all of the COs he knew. Some of Beneze's coworkers relied on drugs and alcohol to mitigate the stress of the job, he said. Others had exhausted these coping mechanisms. Not long before we met, his friend Leonard, another CO, "blew his brains out," he told me. "I don't know the stats, but I've had a lot of friends that killed themselves," he said.

The person who'd introduced me to Beneze was Caterina Spinaris, a therapist from Florence, Colorado, which, like Cañon City, sits at the foothills of the Rockies, in a rugged canyon of steep cliffs and red-rock gorges that straddle the banks of the Arkansas River. Originally from Denver, where she ran a practice that focused on treating the victims of trauma and sexual assault, Spinaris moved to Florence in 2000 to hike, plant a garden, and enjoy the breathtaking scenery. It took her a while to notice the brown and gray detention facilities dotting the landscape, many of them tucked behind fences off unmarked roads that twisted into the hills. Spinaris became aware of their existence when she started fielding an unusual volume of calls requesting referrals from the wives and girlfriends of corrections officers.

The callers were seeking help for their partners. Having never set foot in a prison, Spinaris searched about for a local agency to which she could refer them. When she discovered there was no such agency, she decided to open one herself. Desert Waters Correctional Outreach began with a hotline and an email address that was soon flooded with messages from prison guards who unburdened themselves of their bitterness and anger, which was spilling into their relationships and seeping into their family lives. "I can't seem to get along with anyone anymore," one guard wrote to Spinaris. "I wish I could get out of this rut. That is what prison work is—a big, deep rut." Another wrote, "After a while and numerous incidents, you have so many Band-Aids on you that inmates can't penetrate them and get to you or your 'old' heart. The only problem is the Band-Aids don't come off after work. They stay on. So you live your life and miss all the beauty."

Spinaris heard from guards who were alcoholics, from guards whose marriages were imploding, from guards who seethed at performing a public service that no one ever thanked you for doing and that plenty of people looked down on. The outpouring of unfiltered anguish reminded her of her sessions with trauma victims. In 2012, Spinaris distributed a questionnaire to more than three thousand correctional workers across the country that was designed to measure the prevalence of PTSD. Thirty-four percent of the COs who responded to the survey reported symptoms, a rate comparable to that in the military. In Spinaris's view, many prison guards struggled with something else—feelings of alienation and shame that churned beneath their gruff exteriors.

Spinaris described corrections officers as "an invisible population." This was certainly true of the scholarly literature on prisons, a body of work that focused overwhelmingly on the plight of incarcerated people, perhaps because its authors viewed guards less sympathetically. One exception was a book titled *Prison Officers and Their World*, by Kelsey Kauffman, a sociologist who, after working at a women's prison in Connecticut, spent several years interviewing corrections officers in Massachusetts. As her study showed, violent prison systems inevitably dehumanized not only the prisoners but also the guards. "You can't be a positive person in a place like this," one CO told her. "I used to have a lot of compassion for people and now I don't have as much," said another. "It just doesn't bother me the way it used to, and it bothers me that it doesn't bother me."

In Kauffman's view, the callousness of COs was often preceded by a period of adjustment as guards struggled to negotiate "discrepancies between their own ethical standards and the behavior expected of them as officers." For those who relished violence, the adjustment was easy enough. But this was not typical, Kauffman found. "Initially, many attempted to avoid engaging in behavior injurious to inmates by refusing (openly or surreptitiously) to carry out certain duties and by displacing their aggressions onto others outside the prison or themselves," she wrote.

> As their involvement in the prison world grew, and their ability to abstain from morally questionable actions within the prison declined, they attempted to neutralize their own feelings of guilt by regarding prisons as separate moral realms or by viewing inmates as individuals outside the protection of moral laws . . . Regardless of whether officers became active participants in the worst abuses of the prisons or were merely passive observers of it, the moral compromises involved exacted a substantial toll.

Kaufmann called guards "the other prisoners."

For Black prison guards, an added moral tension existed—the discomfort of working in a system that disproportionately harmed their own communities. This fact was not lost on Black prisoners, who sometimes heckled them for betraying their brothers or working for the Man. In recent years, some Black police officers have heard similar accusations at protests organized by the Black Lives Matter movement where demonstrators have labeled them "sellouts" and "Uncle Toms," insults that carry a particular sting for those who sympathize with some of the movement's aims. A Black CO from Florida whom I'll call James (he did not want his real name used) told me he'd heard his share of such accusations from prisoners through the years. He insisted they didn't bother him, telling me that whenever it happened, he would remind them that he came from a similar background and that they were behind bars because they'd violated the law. "I tell them, that's why you're on that side," he said. Every so often, though, James would be reminded how unevenly the law was applied. One time when he was driving, an officer pulled him over, placed him in handcuffs, shoved him into the back of a police car, and called headquarters to run a background check on the "asshole" he'd picked up. To prepare for such encounters, which happened frequently, James made sure his badge was readily accessible, but flashing it didn't always help, a problem he attributed to the fact that many cops "feel corrections officers are not *real* officers" as well as to his race. "A lot of times they won't recognize me as a law enforcement officer; they just look at me as being Black," he said. In fact, this was a struggle not only with cops but also with his peers in the DOC. "I don't like you, 'cause you're Black," some of his fellow COs told him. "I'm Klan," he'd heard others say. A sixteen-year veteran, James saw his career nearly end when a couple of white guards tried to forge his signature onto a disciplinary report that contained spurious allegations about a prisoner (officers who wrote up false DRs could lose their certification). The officers were acting at the behest of a rabidly racist colonel who had it in for him, James was convinced. Every Black prison guard in Florida had such stories, he told me, compounding the stress of what was already a taxing job.

"THE LOWEST-RANKING MAN"

Perhaps because he'd entered the profession late, Bill Curtis, who had retired three months before we met, emerged from his experience relatively unscathed. Clad in shorts and flip-flops, he'd just returned from a cycling trip to Monument Valley, sporting a newly grown-out beard and a weathered suntan. We had lunch in the outdoor section of a Mexican restaurant on a scorching day in June, drinking pints of Fat Tire to temper the sultry weather and the Cajun burgers we'd ordered. Curtis seemed in good spirits, showing me pictures of the catamaran on which he took his grandkids whenever they came to visit (sailing was another of his hobbies). At no point did he peer nervously over his shoulder or appear to be watching his back. But there was one occupational hazard he told me he hadn't managed to escape, a shift in values and outlook that he believed all corrections officers underwent.

"Your morals change," he said. "It's a coarsening and a lessening of concern for people. It's a slide. When a man—a good man, or woman—goes into prison, a little bit of your goodness wears off. You become jaded. You become more callous. Your language and your interpretation of things changes."

Among the things that led Curtis to feel jaded was his understanding of who took the blame—and who did not—when egregious abuses took place. When he started out at CCI, Curtis assumed that guards who engaged in misconduct would be weeded out. "The 'bad actors' are known and their days as correctional officers are numbered," he wrote in his journal. As he grew more experienced, he began to realize that brutality was routinely excused and not infrequently rewarded. The officer who doused a prisoner in the face with bleach, for example, was not disciplined. He was promoted to sergeant. "Cover-ups, false statements, coercion, and outright lying seem to be the order of the day in this business," Curtis noted with dismay in his journal.

Curtis's cynicism on this score only deepened when, in 2012, after he'd been working for eight years as a CO, the Teamsters hired him to represent officers at different prisons who stood accused of engaging in disciplinary infractions. On occasion, the guards in question had behaved improperly. More often, they were scapegoats who had been singled out to deflect blame from their superiors—the "good old boys" who ran the system and protected the people who displayed loyalty to them. The officer disciplined or demoted was "always the lowest-ranking man," Curtis told me, and often the *least* corrupt worker on duty. A case in point was the death of Richard Mair, a prisoner in the Transitional Care Unit at Dade CI who hung himself in September 2013, one year after Darren Rainey died. A note found in Mair's boxers, titled "FUCK THE WORLD," detailed the abuse that precipitated his suicide, including an incident when a guard ordered Mair, a rape victim, to "strip out" and then, promising cigarettes in return, commanded, "Stick a finger in your hole." In his suicide note, Mair indicated that he tried to file a grievance but that a lieutenant intercepted it, slamming him against a wall and warning him "to keep my mouth shut or else." Two low-ranking guards were subsequently faulted, purportedly for failing to conduct timely checks of Mair's cell before he hung himself. According to Curtis, the real reason they were blamed was for collecting written documentation about what happened that another prisoner, Damien Foster, had compiled and bringing it to the attention of an assistant warden. By the time an investigation was launched, the documentation had conveniently disappeared, Curtis told me, and Damien Foster had been transferred to another prison, preventing him from talking to investigators. "It was a cover-up," Curtis said. At some prisons, another CO told me, corrupt captains and colonels would groom low-ranking guards to do their "dirty work"—roughing people up, writing up bogus DRs—and then pretend to know nothing about it when the conduct was exposed and someone needed to take the fall.

In prisons as elsewhere, dirty workers performed another essential function, shouldering the blame for inhumane systems within which they ultimately had little power and thus deflecting attention from other social actors with far more sway. These other actors included not only their superiors but also judges, prosecutors, and elected officials operating with broad consent from the public. The corrections officers at places like Dade were the agents of a society that was home to the world's largest prison system, a system that grew even faster in Florida than in the United States as a whole, under Democrats and Republicans alike. In 1993, the year before Bill Clinton signed a punitive new crime bill, a series of headline-making murders took place in Florida. The victims were European tourists, which threatened to tarnish the state's image as a family-friendly travel destination. Against this backdrop, Lawton Chiles, Florida's Democratic governor, unveiled the "Safe Streets" program, which called for building twenty-one thousand new prison beds. "It's time to stop talking tough about the problem and start acting tough," Chiles declared. Five years later, Jeb Bush, Chiles's Republican successor, signed the "10-20-Life" law, which mandated a minimum ten-year sentence for anyone armed with a gun during certain felonies, irrespective of the circumstances. Florida's legislature later extended the law to apply to sixteen- and seventeen-year-olds. Between 1970 and 2010, Florida's prison population grew by more than *1,000* percent. It grew when the crime rate was rising and when it flattened out—year after year, legislative session after legislative session. Someone needed to do the dirty work of running this system on a day-to-day basis, and someone needed to foot the blame when its brutal inner workings spilled into the headlines on occasion, prompting "good people" to express dismay and shock.

The fact that the blame fell on low-ranking workers, and not on the respectable people who relied on and benefited from what they did, was nothing new. In antebellum America, a similar logic shaped popular attitudes toward another band of dirty workers: the auctioneers and traffickers who presided over the interstate slave trade, here parading their "merchandise" in the showrooms of cities like New Orleans, there dragging chained caravans of slaves through the streets of Washington, D.C. These traders played a crucial role in enabling slavery to spread and thrive, particularly after 1808, when the ban on importing slaves from Africa led southern planters to turn to domestic sources to replenish their supply of "field hands." Soon enough, slave traders were subjected to withering scorn, not only in the pages of abolitionist journals but also, tellingly, in much of the South. In his popular 1860 book, *Social Relations in Our Southern States*, Daniel Hundley, the son of a plantation owner from Alabama, denounced slave traders as "soul drivers" who plied a "detestable" trade. "The miserly Negro trader . . . is not troubled evidently with a conscience, for, although he habitually separates parent from child, brother from sister, and husband from wife, he is yet one of the jolliest dogs alive," wrote Hundley, whose book nonetheless offered a robust defense of slavery. On the floor of Congress, the first person to denounce the interstate slave trade was Virginia's John Randolph. Like Hundley, Randolph was a defender of slavery who felt moved to say something after a high-ranking foreign visitor told him that he was "horrorstruck and disgusted" by the sight of coffles of slaves passing through the streets in broad daylight.

Denouncing slave traders enabled southerners like Randolph to shame the culprits responsible for the horror while leaving the institution of slavery unquestioned and absolving themselves. Whereas traders were greedy opportunists driven by profit, slave owners were men of honor. Whereas traders destroyed slave families, masters took pains to protect them. Although these distinctions were spurious, drawing them was not an empty rhetorical exercise. It was "therapeutic," notes the historian Robert

Gudmestad, enabling southerners to distance themselves from what Gudmestad calls "a troublesome commerce." Troublesome not because of the torment it inflicted on Blacks, but because of the uncomfortable truth it threatened to bring home to whites: that slavery itself was detestable; that it routinely tore families apart. Troublesome, too, because the trade's most repugnant features were embarrassingly public.

The disreputable "soul drivers" who trafficked in slaves served as convenient scapegoats for slavery's champions and apologists. And yet, as some shrewd observers noted at the time, not all of them were seen as disreputable. As the abolitionist Theodore Dwight Weld pointed out, the stigma of the slave trade fell mainly on the shoulders of "men from low families" who carried out the "vulgar drudgery" that earned their vocation its notoriety. The largest, most successful slave traders were spared such opprobrium. "There was apparently little stigma attached to the trade for those who were successful at it," observes the historian Walter Johnson in his magisterial study of a slave market in New Orleans, *Soul by Soul*. Not coincidentally, the wealthy men who ran the largest trading firms tended to outsource much of the actual work to roving bands of lower-class laborers, from whom they made sure to distance themselves in more polite company. "See those gentlemen, I have nothing to do with that," remarked one successful trader when asked about a slave sold in his name, neglecting to mention that the "soul drivers" in question were working for him.

In the spring of 2020, prison guards in America received a new designation, joining the ranks of "essential workers"—truck drivers, warehouse handlers, grocery clerks—who were instructed to continue showing up at their jobs even as the coronavirus pandemic led mayors and governors to order lockdowns. Some of the frontline workers who did these jobs—most notably, physicians and first responders—were soon accorded hero status for the risks they braved and the sacrifices they made. In New York, where the first wave of the pandemic took a particularly devastating toll, citizens stood on their stoops and balconies every evening to salute the medical professionals scrambling to accommodate the influx of COVID-19 patients pouring into the city's hospitals. In the course of trying to provide ventilators and some measure of comfort to these patients, many medical workers themselves fell ill, which only increased the public's gratitude to them.

Nobody clapped for corrections officers, whose risk of infection was equally grave. The crowded, unsanitary conditions in America's prisons was one reason for this. The lack of regard for the health and welfare of the people in them—both the prisoners doing time and the staff watching over them—was another. In New York City alone, more than twelve hundred guards tested positive for the coronavirus, and thirteen prison staff died during the pandemic's initial wave. Some COs later complained that they were pressured to return to their jobs while they were still symptomatic. Others alleged that they were actively discouraged from wearing masks by their supervisors, a problem not confined to New York. In Florida, the *Orlando Sentinel* interviewed guards at four different prisons who came to work with their own masks, only to be reprimanded by their superiors. A report on the federal prison system published by the Marshall Project uncovered multiple instances in which staff alleged that they lacked protective gear and were pressured to continue working even after exposure to the virus.

Over the course of 2020, nearly 100,000 corrections workers tested positive for the virus and 170 died. The COs losing their lives looked a lot like the casualties in other frontline jobs—people like Quinsey Simpson, an African American man from Queens who developed a cough after doing a shift in a security booth at Rikers, unaware that the guard he'd replaced was symptomatic. Simpson, who had not been supplied with gloves or a mask, soon developed respiratory problems. He died shortly thereafter, leaving behind a six-year-old son.

If prison guards were not showered with applause for continuing to do their jobs, it's perhaps because, by the time the pandemic began, public attitudes about the crowded, violent facilities in which they worked had shifted. After decades of backing punitive laws and harsh sentencing policies, Americans in many parts of the country had begun to embrace reforming the criminal justice system. In 2019, the New York City Council approved a plan to close Rikers Island by 2026, an idea that would have been unthinkable a decade earlier. But if treating prisoners more humanely was a public priority, you wouldn't have known it from the way prosecutors and elected officials carried on during the pandemic. By June 2020, all five of the nation's largest COVID-19 outbreaks were in correctional institutions. At a prison in Ohio where dormitories were filled to double their capacity, nearly three-fourths of the people in custody were infected. In the face of these alarming figures, some advocates and public defenders urged elected officials to discharge low-level offenders and older prisoners. Governors in a few states responded by releasing thousands of incarcerated people who were nearing the end of their terms. But in many other states, little was done. "Despite all of the information, voices calling for action, and the obvious need, state responses ranged from disorganized or ineffective, at best, to callously nonexistent at worst," a survey conducted by the Prison Policy Initiative and the ACLU found.

By early 2021, the jail population in many states had returned to pre-pandemic levels, even as the number of infections continued to rise. In states like North Carolina and Wisconsin, the virus's toll on correctional workers prompted officials not to rethink the logic of mass incarceration and put fewer people behind bars, but to shut down understaffed facilities and transfer prisoners elsewhere. The closures exacerbated overcrowding in the penitentiaries that remained open—and, in turn, the fear of infection among both prisoners and staff. "They're terrified," an official with the union representing prison guards in North Carolina told *The New York Times*. "They feel like, as usual, they're forgotten and left behind."

Feeling forgotten was a familiar sensation to a CO I'll call Bobby who worked at CCI, the same prison where Bill Curtis had cut his teeth as a guard. Like Dade, CCI attracted some media attention after a prisoner there died. The death occurred late one night, when guards conducting a "spot" compliance check roused a man named Matthew Walker from his sleep and ordered him to put away an item that they claimed was out of place. The misplaced item was a plastic cup. "This is crazy . . . you are waking me up about a cup!" Walker fumed, prompting a confrontation that soon erupted into a bloody melee. By the time it was over, two officers had been injured, and Walker had sustained "blunt trauma" injuries in eleven places, including to his neck and head. At 1:20 a.m., he was pronounced dead, a death that was "tragic, senseless and avoidable," a grand jury report subsequently concluded.

The grand jury report presented conflicting evidence about what had caused the skirmish to escalate. But it was clear on one point, which was that conducting late-night cell-compliance checks, a policy devised by a captain at the prison, was a "bad idea" that numerous guards feared would spark violence. Afterward, nine COs at CCI were dismissed, a move that Bill Curtis had described to me as a public relations exercise: Florida's governor, Rick Scott, was running for reelection and, in the wake of the *Herald* story about Darren Rainey, wanted to show "how tough he is on inmate mistreatment." The toughness had limits, Curtis pointed out. While the guards were punished, the warden at CCI was promoted to regional director. An assistant warden became full warden. The captain who initiated the compliance checks asked to be transferred to another facility, a request that was obliged.

One of the guards who was disciplined was Bobby. A compactly built man with a steely gaze and a Home Depot hat pulled low over his brow, Bobby described the cell-compliance policy at CCI as "a ticking time bomb" that was bound to provoke a violent reaction. "Everyone said someone's gonna get hurt," he told me. Yet no subordinate could apparently say this without fearing retaliation. "Do it or I'll replace you" was the message sent to line staff, Bobby said. Bobby wasn't present when the skirmish with Walker began—he rushed over afterward to provide backup—or at the end, when evidence he believed should have been preserved was removed or mishandled (a point the grand jury report also raised). This didn't prevent him from losing his job.

I wanted to know what had led Bobby to become a corrections officer in the first place. He needed a job, he said, and although the salary was lousy, it came with benefits. Better wages typically meant no benefits, and Bobby had a family to support. "You either have the benefits and no pay, or you have pay but no benefits" was how Bobby summarized his employment options. In the decades after World War II, workers in America's mills and factories often managed to secure both of these things—decent pay *and* benefits—but that era was long gone, and Bobby knew it, so he swallowed his pride and took what he could get, a dangerous, low-prestige job that ended in dishonor and disgrace.

Or rather, that would have ended there, if not for the fact that later, after Governor Scott was reelected and attention shifted elsewhere, all of the guards at CCI who were dismissed were reinstated. The move prompted some newspapers in Florida to express outrage at the lack of accountability within the DOC. The outrage was justified. According to the grand jury report, after Matthew Walker was beaten, a lieutenant involved in the melee stood over him and yelled, "Do you know who I am? I will kill you, motherfucker!" By the time medical help finally arrived, Walker was no longer breathing. As with the murder of Darren Rainey, the fact that nobody paid a lasting price for Walker's death showed how little value was placed on incarcerated people's lives. Still, one could hardly blame Bobby for wondering why the media's outrage wasn't directed at the senior officers who had instituted the policy that precipitated the incident rather than at the low-ranking guards. If the public really cared about abuse, he added, Florida would "increase the pay of officers to where people would actually want a career out of it" rather than making them the fall guys for a corrupt system that made good officers feel devalued.

Were there places where prison guards *didn't* feel devalued? In *Prison Worlds*, an ethnographic study of the prison system in France, the anthropologist Didier Fassin found that many guards were so embarrassed about what they did for a living that they avoided talking about it. "I never tell anyone what I do . . . it's too shaming," one guard said. "I don't tell my friends," another confessed. Most of the guards in Fassin's study hailed from small provincial towns and working-class families. Many compared themselves unfavorably to police officers, sensing that the latter regarded them as "sub-professionals." Far from taking pride in the uniforms they wore, the guards felt "contaminated by the conditions in which they carry out their work," Fassin found. In other words, like their counterparts in places like Florida and Colorado, they felt like dirty workers who performed a demeaning and morally discredited job, notwithstanding the fact that the conditions they worked under reflected the public's wishes and priorities. (Another study of prison guards in France asked "why the image of the correctional officer is so negative, when all he does is carry out the task society sets him," suggesting that the answer could be traced to "the process of displacement," whereby society's "bad conscience" about the deplorable conditions in France's prisons were projected onto the figure of the guard.)

But while such feelings were pervasive, they were not inevitable. In 2015, the journalist Jessica Benko toured the grounds of Halden Fengsel, a maximum-security prison in Norway. Located in a bucolic landscape of pine trees and blueberry bushes, the prison housed 251 people, more than half of whom had been convicted of violent crimes. This did not prevent Halden's proprietors from allowing them to live in furnished rooms with no bars on the doors. The prisoners at Halden were encouraged to attend classes and permitted to circulate unmonitored by surveillance cameras, a philosophy known as "dynamic security," which sought to reduce violence through trust and social interactions rather than coercion and control. Halden had only one isolation cell reserved for unruly individuals, equipped with a restraining chair that had never been used. A skeptic might note that such a prison would probably cost more to run than a maximum-security prison in the United States, which was true, and that the data on whether more humane prisons succeeded in reducing the recidivism rate was inconclusive. But reducing the recidivism rate was not the only goal. Equally important was creating an institution in which Norwegians could take pride, a sentiment the staff at Halden appeared to share. "I have the best job in the world," the warden of the prison told Benko, mentioning that his officers liked their jobs and hoped to finish their careers there. As the comment suggested, it wasn't only the dignity of incarcerated people that was at stake in changing the brutal conditions in America's prisons. It was also the dignity of the staff, who didn't use fear and threats to enforce their authority and didn't seem to feel contaminated by the conditions in which they worked.

At the very beginning, Bobby actually did take pride in his job. He wanted to make a career of it, he told me. The low pay and rampant corruption made this impossible, he'd concluded, conditions he wasn't holding his breath for "good people" in Florida to change. The same critics who rushed to blame the guards at CCI for killing Matthew Walker after hearing about it on the local news would just as quickly forget that the prison even existed, he predicted, likening the level of awareness about the conditions in Florida's prisons to the level of awareness of the landfills where the public's trash was dumped. "You put your trash out, the trash gets taken away—you don't care about the landfill," he said. "The only time you care is when it's full and you gotta pay for a new landfill." For most Floridians, it was the same with the DOC, he told me at the restaurant where we met, which had begun to fill up for happy hour, the customers trickling in for beers and margaritas served at a bar flanked by TVs tuned to golf and baseball. "Nobody cares about the DOC until something hits the newspaper," Bobby said, tugging on the brim of his Home Depot cap as he glanced over at the scrum of patrons by the bar, "and then the media's gonna blow it up to make it sound like, you know, we're a terrible group of people."

3

Civilized Punishment

As in most states, prisons in Florida are not easy to access. They're not even easy to find. Without exception, the DOC facilities that I visited were situated in remote areas, off backwoods roads, in fenced-off compounds surrounded by woodlands and swamps. You had to drive a good distance to reach their entrances, past the beaches and theme parks, the golf courses and vacation resorts, the palm-lined boulevards and ocean-front condos that drew throngs of tourists and snowbirds to the Sunshine State every year. The closer you got to the gates of a prison, the more the traffic thinned out, and the spottier the cellular coverage generally got. On several occasions, I zoomed past the facility I'd set out for, only to notice later that my GPS had stopped working and that I'd gone too far. Other than marshes and fields and some unwelcoming signs—UNANNOUNCED PERSONS WILL BE PROSECUTED—there was little else around.

Along with their remoteness, the prisons I visited were notable for their drab architectural style: featureless, box-shaped buildings arrayed across sterile compounds that projected a bland orderliness. Squinting at them through rows of security fencing and the harsh glare of the Florida sun, I rarely saw much in the way of human activity: the rec yards were empty, the grounds still. Aside from the occasional squawk of a bird circling overhead or the buzz of a mosquito that landed on my arm, I heard few sounds. Nothing eventful was happening here, the dull facades of the generic-looking buildings suggested, certainly nothing disturbing or violent.

Prisons have always been geographically isolated and visually lackluster,

I assumed after these visits, the better to avoid attracting unwanted attention from outsiders. As I subsequently learned, this assumption was wrong. During the Jacksonian era, "Americans took enormous pride in their prisons, were eager to show them off to European visitors, and boasted that the United States had ushered in a new era in the history of crime and punishment," notes the historian David Rothman. Among the visitors invited to see the early prisons of the new republic was Charles Dickens, who toured the grounds of Pennsylvania's Eastern State Penitentiary in 1842, chatting freely with convicts as he passed from cell to cell. "Nothing was concealed or hidden from my view," Dickens wrote in his *American Notes*, "and every piece of information that I sought, was openly and frankly given." In both America and England, prisons in the nineteenth century tended to be built in prominent places that were exposed to the public. Some boasted soaring turrets and stone arcades and were likened to palaces.

As the sociologist John Pratt has documented, it was only later that the architectural style grew more spartan and that prisons started to be built on the "unobtrusive margins" of society. Why did this happen? One explanation is that prison administrators learned from experience not to open their doors to observers like Dickens, who praised the officials in charge of the Eastern State Penitentiary for their good intentions but lambasted the system of "Prison Discipline" they had devised, which confined people to total isolation. The regimen of solitude would cultivate introspection and self-discipline, reformers of the Jacksonian era believed. Dickens was unpersuaded, describing solitary confinement as a "dreadful punishment" that was all the more insidious because its ravages were cloaked and camouflaged. "I hold this slow and daily tampering with the mysteries of the brain to be immeasurably worse than any torture of the body . . . because its ghastly signs and tokens are not so palpable to the eye and sense of touch as scars upon the flesh," Dickens wrote. At the time, America's penal system was shaped by a belief that prisons could be designed to foster moral uplift and turn chastened offenders into law-abiding citizens. By the 1980s, a more punitive philosophy had taken hold, which made prison administrators and public officials all the more inclined to limit access to their grounds.

But the shift to the margins of society could also be attributed to something else. In Pratt's view, it reflected the triumph of "civilized punishment"—civilized not in the conventional meaning of the term, but in the sense that Norbert Elias described, whereby distasteful and disturbing events were removed from sight and pushed "behind the scenes of social life." At first glance, Elias's work appears to bear little relevance to the study of punishment. His two-volume 1939 work, *The Civilizing Process*, a study of European manners that traced the evolution of table etiquette and other behavioral norms from the sixteenth to the nineteenth century, scarcely mentions the subject. But in recent years, scholars of crime and punishment have drawn on Elias's insights to explain some of the ironies and contradictions of contemporary penal practices. At the core of the "civilizing process" was the rise of internal inhibitions that led social actors to suppress the more "animal" aspects of human conduct and to hide such behavior from others, Elias argued. Bodily functions (spitting, farting) came to be seen as offensive and to be banished from polite company. "Disturbing events" such as the carving of dead animals—an activity once performed at the table before festive meals—were hidden in deference to the rising "threshold of repugnance" among members of the courtly upper class.

Though Elias did not examine how this shift in sentiments might have altered the landscape of punishment, his argument that concealment was a central part of the civilizing process seemed strikingly relevant to another set of "disturbing events": the torture and execution of criminal offenders. In medieval and Renaissance Europe, crowds routinely gathered to watch wrongdoers get marched to the gallows, where they were mutilated, burned, and hanged. During the nineteenth century, these so-called spectacles of punishment grew increasingly rare, and many of the practices that had long been theatrically displayed (floggings, beheadings) were outlawed, not least because elites had come to regard them as repellent. In *Discipline and Punish*, Michel Foucault argued that the transition to the more refined technologies of punishment in the modern era—most notably, the prison—was driven by the desire to control and observe the bodies of criminals, rendering them docile and obedient. Criminologists influenced by Elias have emphasized another rationale, arguing that the shift was propelled by a desire to *hide* these bodies from respectable citizens who no longer wanted to glimpse the sordid business of punishment with their own eyes.

The fact that corporal punishment came to be viewed as sordid was, in theory, a sign of progress. Yet as Elias's disciples have noted, the "civilizing process" he outlined did not suggest that brutal violence would cease, only that it would be relegated to more private spaces. According to the scholar David Garland, who introduced Elias's work to the sociology of punishment, violence would not offend civilized sensibilities so long as it unfolded behind closed doors or could be sanitized. (The state's monopolization of violence was a major theme in *The Civilizing Process*, a development that coincided with an emphasis on self-restraint that deterred ordinary citizens from engaging in unlicensed displays of aggression.) In his book *Punishment and Modern Society*, Garland presented a genealogy of "civilized sanctions" that bore out this theory. Hanging criminals on the scaffold was uncivilized, everyone in contemporary America agreed. But executing them by lethal injection—a more discreet method of killing—was legal in many states. Flogging prisoners clearly violated the "threshold of repugnance" among modern Americans. But caging them in hidden, segregated "isolation units" did not. The fact that solitary confinement's "ghastly signs and tokens are not so palpable to the eye," as Dickens had observed in 1842, was precisely why so many people failed to find it offensive. "Routine violence and suffering can be tolerated on condition that it is discreet, disguised, or somehow removed from view," observed Garland. What mattered was not the level of brutality, but its visibility and form. Viewed in this light, the remoteness of Florida's prisons was not an accident. Throughout the Western world, "the civilized prison became the invisible prison," Pratt observed, hiding the system's violence and making it that much easier for "good people" to ignore or forget about what was happening behind the walls.

In fairness, not everyone in Florida put this out of mind. Between visits to penitentiaries, I made my way to a coffee shop one day to meet Judy Thompson. An African American woman from Baymeadows, a suburb of Jacksonville, Thompson was the former coordinator of a leadership academy at the Mayport Naval Station. She was also the founder of an organization that advocated fairer sentencing policies and more humane conditions in Florida's prisons. Thompson came to our meeting with a stack of letters that had been sent to her by incarcerated people who'd heard about her organization and had written to relay stories about the mistreatment they'd endured. The letters arrived faster than she could open them, she said. The ones she'd brought to show me had all been sent to her in the previous month.

Composed by strangers, the letters touched a nerve in Thompson, who had six sons. In 1999, the second youngest of them, Chris, was arrested for participating in a robbery. No one was hurt during the crime, Thompson told me, and Chris, who was twenty-one, had not been armed. Even so, he was sentenced to thirty years. Until that moment, Thompson hadn't thought much about the punitive laws that had turned prisons into one of Florida's largest growth industries. "It just wasn't part of my world," she said. Afterward, she started reading about the wave of punitive measures that had caused the prison population to surge both in Florida and in the rest of the country. Among the books she later came across was *The New Jim Crow*, by Michelle Alexander, a law professor who depicted the war on drugs and other supposedly color-blind policies (mandatory minimum sentences, stop-and-frisk programs) as pillars in a racial caste system that was as pervasive and pernicious as the segregation laws to which Blacks had been subjected before the civil rights movement. *The New Jim Crow* highlighted the devastating costs, becoming a runaway bestseller and re-shaping the debate about mass incarceration. Judy Thompson devoured the book, which appeared in 2010. A year later, she launched her organization, which she called the Forgotten Majority, because, she said, "those behind the walls are truly forgotten" and because, in some cities, a majority of young Black men were ensnared in the criminal justice system. In 2019, Blacks made up 17 percent of Florida's population but nearly half of the state's prisoners (Latinos comprised an additional 40 percent).

Thompson was under no illusions that changing the system in which her son was penned would be easy. Yet she soon discovered that doubts about mass incarceration were emerging in some surprising places. Not long after she formed the Forgotten Majority, she paid a visit to Greg Evers, who chaired the Criminal Justice Committee in the Florida Senate. Senator Evers was a conservative Republican from the Panhandle with an A-plus rating from the National Rifle Association. Thompson went to see him to tell him about a petition she was circulating to make it easier for felons who she felt deserved a second chance to be eligible for parole, which Florida had effectively abolished in 1983 by enacting a truth-in-sentencing law that required all offenders to serve 85 percent of their prison terms. Among the prisoners she believed should be granted relief from this law was her son Chris, who had used his time in prison to become an accomplished jailhouse lawyer, dispensing legal advice to other incarcerated people. (They called him "little Johnnie Cochran," she told me with pride.)

Senator Evers received Thompson in his office. As she made her pitch, he nodded. Then he told her, "Judy, you're preaching to the choir. It's costing us too much money to lock all these people up." Across the country, conservatives who'd spent decades championing law-and-order policies were coming to a similar realization, unsettled by the fiscal costs of building the world's largest prison system, if not by the moral costs. A year later, Thompson attended an event at the Governor's Mansion, in Tallahassee, to honor Black History Month. Before leaving, she spotted Florida's governor, Rick Scott. She walked up to him, told him about her organization, and said that she found it impossible to enjoy Black History Month when so many Black men were languishing in Florida's prisons "without hope." Rather than take offense, Scott, a Republican who in a few years would become a vocal supporter of Donald Trump, invited her to set up a meeting. Afterward, Thompson assumed that she would never hear from him again. Several weeks later, her phone rang. It was an aide to Governor Scott, calling to follow up. On her next trip to the Governor's Mansion, Thompson hauled along several garbage bags filled with signed petitions calling for a restoration of parole. Changing the law was up to the legislature, Governor Scott told her, but Thompson left the meeting heartened that he'd listened to her.

At long last, policy makers were beginning to reckon with the calamitous effects of mass incarceration, it appeared, not only in Florida but across the country, where liberals and conservatives started to forge alliances to promote less punitive sentencing policies that could reduce the prison population. In the years to come, numerous states would curtail or eliminate mandatory minimum statutes and give judges greater discretion over how to sentence convicted felons. In 2016, Florida repealed the "10-20-Life" law, which had imposed draconian sentences on anyone armed with a gun during a crime. But persuading politicians like Rick Scott to put fewer people in prison, which cost between twenty thousand and thirty thousand dollars a year, was one thing. A less punitive approach could save taxpayers money. Making the case that prisoners should be treated more humanely—by hiring better-trained staff, by providing adequate mental health services, by increasing oversight and monitoring, all of which stood to *cost* money—was another matter entirely. As James Forman, a law professor at Yale, pointed out, critics of mass incarceration often focused on the ravages of the drug war and other policies that drove up the number of nonviolent offenders behind bars. Missing from this indictment was the fact that even if all nonviolent drug offenders were released, America would *still* have the world's largest prison system—a system that stood out as much for its brutality as for its size.

When she formed the Forgotten Majority, the brutality of Florida's prisons was not high on Judy Thompson's agenda, not least because she had no idea how brutal they were. Then she started receiving letters like the ones that she spread on the table at the coffee shop where we met. In one of them, a prisoner described going to a captain to seek protection from gang members who had threatened him. The captain responded by telling him that if he did not shut up, "he would spray me (pepper spray) and . . . knock my teeth out," the prisoner wrote. I'd contacted Thompson because I wanted to know how unusual the abuses at Dade were. Although Dade was an extreme case, physical and verbal abuse was not unusual at all, she told me. Most of the letters she got were from prisons in northern Florida, she said. In Thompson's view, it was not a coincidence that the farther north one went, the less ethnically diverse the staff became. At a penitentiary near Jacksonville, two COs were convicted of plotting to kill a Black prisoner. The guards were members of the Ku Klux Klan.

For years, the inhumanity of Florida's prisons attracted little notice. But in the wake of Darren Rainey's murder and the wretched conditions described in the *Herald* and other news outlets, a small group of advocates started pushing for change. One of these advocates was George Mallinckrodt, the psychotherapist who had gotten fired after voicing complaints about the abuses at Dade. Another was Thompson, who, in March 2015, attended a legislative hearing in Tallahassee, chaired by Senator Evers, to address the rampant, sometimes-lethal violence in Florida's prisons and jails. As usual, Thompson came to the hearing with some prisoners' letters, including one she had received two years earlier from Mark Joiner, a prisoner who had been at Dade and who had contacted her a few months after Darren Rainey died. In his letter, Joiner described the sadistic abuses meted out to patients in the prison's Transitional Care Unit, including the "scalding hot water" that had been used to torture Rainey and other mentally ill people. When Thompson first read the letter, she didn't believe it, she told me, because the allegations were "too heinous." Then Joiner wrote again, repeating the same charges, at which point she realized he was telling the truth. One section of Joiner's letter had particularly moved Thompson, a passage in which he compared the impunity granted to the guards who killed Rainey to the life sentence he'd received for his crime (Joiner had been convicted of murder). At the hearing in Tallahassee, Thompson decided to read this section aloud. "I've come to the conclusion that one is prosecuted for murder depending on who you are," Joiner wrote. While some murderers were "committed to DOC for life," the letter continued,

> *others collect a paycheck from same and go home daily, free to continue the behavior at another time. One day, I wonder if the people of this state and the nation will look back at this time . . . and say, How could this go on? Law enforcement knew this, representatives in state government were aware of it, yet it just continued on.*

After Thompson finished reading, some corrections officers also testified about the untenable conditions in Florida's prisons—the overcrowding, the understaffing, the low morale. Their words made an impression on some of the lawmakers on hand, among them Senator Evers, who spent the rest of the spring promoting a sweeping prison reform bill. The bill included provisions to bolster staffing at prisons and to increase penalties for officers who engaged in abuse. It also called for establishing an independent oversight commission that would monitor DOC facilities by conducting surprise inspections and identifying problems before they metastasized.

In other Western countries, monitoring prisons in this way was not unusual. In the United Kingdom, for example, there were three separate layers of oversight: the Inspectorate of Prisons, which carried out unannounced inspections of all detention facilities; the Prisons and Probation Ombudsman, which investigated complaints; and boards of local citizens that conducted additional reviews. The goal was "not to play gotcha," said Michele Deitch, an expert on prison management who teaches at the University of Texas at Austin, but to detect problems early. "It's preventive." Creating strong oversight mechanisms also reflected a belief that the public would be better served if, like other large institutions (banks, nuclear facilities), prisons were subject to independent scrutiny. In the United States, a different view prevailed. Prisons were "the ultimate enclosed institutions, completely lacking in transparency—to the public, to the press, to everyone," said Deitch. Instead of preventive mechanisms, the model for remedying problems in the United States, historically, was through court-ordered consent decrees, settlements that resulted from class-action lawsuits filed against states whose jails or prisons were plagued with systemic deficiencies. One shortcoming of this approach was that it invariably came after the fact, "once things have already hit rock bottom," noted Deitch. Another was that the improvements tended to be short-lived. Once officials complied with the court order, attention shifted elsewhere. Then the problems resurfaced, especially in states where prisons were left to monitor themselves.

At the panel hearing chaired by Senator Evers, several officers spoke about why the inspector general's office—which was part of the Florida Department of Corrections—could not be entrusted with oversight. "We're no longer to the point where we can police ourselves adequately," said an officer whose investigations of abuse had been sabotaged. After leaving Tallahassee, Judy Thompson felt optimistic that the question Mark Joiner posed in his letter—how could a civilized country allow such things to go on?—would be answered by affirming that they must *not* go on. Then Thompson watched the Florida House of Representatives present its own reform bill—a bill that stripped out all of the meaningful provisions in Senator Evers's version, including the proposal to create an independent oversight commission, on the grounds that this would give the DOC "a bad tag." In the end, nothing passed, save for a few toothless measures that Governor Scott approved at the last minute through an executive order.

When we met, I asked Thompson how much she thought Florida's leaders cared about the abuse of prisoners in light of all this. She pointed to the latest pile of letters she'd gotten. "Not enough," she said with a sigh, "not enough." Then she added, "If, in fact, Governor Scott feels he doesn't have to take action on this, he's probably right about that, because the average person who's not impacted by incarceration couldn't care less."

"MANDATE BY DEFAULT"

One reason the average person ought to care about incarceration is that it is a collective undertaking—financed by taxpayers, presided over by legislators and judges, audited by inspectors who work for the state. However inaccessible and hidden from view they are, jails and prisons are public institutions. Yet in Florida, as in many other parts of the country, it isn't so simple. Before going into politics, Rick Scott had served as the CEO of Columbia/HCA, a for-profit hospital chain. Along with staff cuts, one of the ways that Scott proposed to lower spending on prisons after he became governor was by contracting out services to the private sector. When Harriet Krzykowski worked at Dade, her employer was not the Florida Department of Corrections. It was Corizon, a company based in Brentwood, Tennessee, and, later, Wexford, which was based in Pittsburgh. For these firms, the sordid business of punishment was just that—a business, and a potentially lucrative one in a state with nearly a hundred thousand people behind bars. In 2012, the same year Darren Rainey was tortured and killed, Wexford and Corizon were awarded five-year contracts worth a combined $1.3 billion to deliver psychiatric and medical care to incarcerated people in Florida.

As I discovered on my visit to Dade, one consequence of privatization was to push what happened inside prisons further behind the scenes, into the hands of companies that felt no obligation to answer for their conduct. I came to the prison at the invitation of the DOC, which after repeated requests finally agreed to let me tour the facility in the company of Glenn Morris, the new assistant warden. When I asked Morris if I could talk to some of the mental health counselors in the TCU, however, he said this decision was not up to him. It was up to Wexford, to which the state had transferred responsibility. From Morris's office, I called a DOC spokesman in Tallahassee who confirmed that I would have to obtain permission from the company. I proceeded to call Wexford, reaching a media relations officer in Pittsburgh who told me that no one on the Wexford staff was available to talk to me. Because I was already at Dade, I would be happy to wait until someone was available, I told the Wexford spokesman. No interview would be granted, he said.

Wexford declined to answer any questions about the abusive practices that took place at Dade. Corizon was a bit more forthcoming. Dr. Calvin B. Johnson, the company's chief medical officer, told me that all of the company's employees went through an orientation that introduced them to a set of "company-wide values" encapsulated by the acronym SMART—Safety, Motivation, Accountability, Respect, and Teamwork. If mental health counselors witnessed unethical conduct, they were given multiple channels to report it, he added, including an anonymous hotline routed to corporate headquarters. This information came as news to Harriet Krzykowski, who told me she had never heard of the motto SMART when she worked at Dade. She was equally unaware of the anonymous hotline. Beyond the warning she received not to antagonize security, she couldn't recall any directives during or after her orientation about what to do if she witnessed abuse. Her observation was echoed by several of her former peers, including George Mallinckrodt, who said there was "never any guidance, any training, any protocol, zero from Corizon, not only in regard to what abuse was but how to report it."

One danger of relying on the private sector to provide mental health services in prisons is that companies may have an incentive to hide problems that might jeopardize their contracts. Another is that the drive to minimize costs could undermine the quality of care. Not long after Florida privatized medical services in its prisons, Pat Beall, a reporter at *The Palm Beach Post*, published an investigative story that detailed how the experiment was playing out. Seven months into the trial, Beall found, the number of deaths in custody across the state had soared to 206—a ten-year high. Among the deceased was a female prisoner who had been given Tylenol and warm compresses for lung cancer. Other prisoners complained that their prescription painkillers were suddenly replaced with cheaper substitutes, like ibuprofen. Meanwhile, the number of prisoners with serious illnesses who had been sent to outside hospitals had fallen dramatically, by 47 percent, which lowered costs but endangered lives. "We order surgery and they don't come in," a doctor told Beall. "They are dying before they get to surgery."

What Beall found in Florida echoed what investigations had turned up in other states. In 2012, a court-ordered investigation found "serious problems with the delivery of medical and mental health care" at a prison in Idaho, where patients under Corizon's care were left in soiled linens, denied access to food and water for days, and given the wrong medications. The report characterized the deficiencies as "frequent, pervasive and longstanding" (Corizon dismissed its findings as "misleading and erroneous"). In Kentucky, the families of several prisoners filed lawsuits after seven people died within a year at a jail where Corizon delivered the care, among them a severe diabetic who developed an infection and was kept at the facility rather than taken to an emergency room.

But privatization didn't merely imperil the health of incarcerated people. It could also compromise the well-being of the staff delivering the care, forcing them to make decisions based on cost considerations rather than on medical need. In 2014, at a jail in Chatham County, Georgia, a man named Matthew Loflin pleaded to be sent to a hospital after experiencing shortness of breath and passing out in his cell. Dr. Charles Pugh, an on-site Corizon physician, asked one of the company's regional medical directors for permission to transfer Loflin. The request was denied. In the days that followed, as Loflin's condition worsened, Dr. Pugh and two Corizon nurses repeatedly asked for permission to send Loflin to the hospital, to no avail. Eventually, Dr. Pugh arranged to have Loflin see a cardiologist, in the hope that *he* could send Loflin to the emergency room. By the time this happened, Loflin's blood pressure had plunged, and he lost consciousness again. Shortly after he was finally admitted to a hospital, he died.

Dr. Pugh's experience was described in "Profits vs. Prisoners," an investigation of Corizon conducted by the Southern Poverty Law Center. The investigation opened with an account of another case, involving a prisoner in Oregon named Kelly Green who broke his neck during a psychotic episode (Green was schizophrenic). Instead of sending Green to the hospital right away, Corizon staffers released him from jail and then drove him to a hospital so that someone else could admit him. This arrangement, known as a "courtesy drop," spared the company from having to pay for the hospital stay, at the expense of incarcerated people whose care was delayed. Like Matt Loflin, Green ended up dying, a death his family was convinced could have been avoided if he had been taken to the hospital immediately (the family ended up suing Corizon, which denied any wrongdoing).

The Southern Poverty Law Center report suggested that, thanks to privatization, nurses, doctors, and psychiatrists were routinely thrust into such roles, pressed to do the dirty work of keeping costs down for companies that were paid a flat rate for their services, meaning that "every dime saved on prisoner care is a dime added to the company's bottom line." In a declaration submitted in court on behalf of the family of Kelly Green, Dr. Pugh described the pressure this generated as unrelenting. "Once or twice a week, there were telephone conferences I was expected to attend with the Corizon regional medical director regarding who was in the hospital and what was going on with patients in the hospital," he wrote. "There was a constant demand to monitor all hospitalizations, to avoid hospitalizations, to request prompt hospital discharges and minimize hospital stays." Some staff members presumably lost little sleep about succumbing to these demands, either because they had become desensitized to the needs of the incarcerated people in their care or because they wanted to avoid antagonizing their superiors. The latter concern was a valid one. After the death of Matt Loflin in Georgia, Dr. Pugh and the two nurses who had requested that he be transferred to a hospital shared their concerns about patient safety with a sheriff. According to the Southern Poverty Law Center investigation, all three were subsequently fired.

"A special burden of accountability accompanies grants of public authority," observes John Donahue in his authoritative study, *The Privatization Decision*, which examines the circumstances under which contracting out public services to for-profit companies is warranted. While privatization can make government more efficient, this is not the only criterion by which it should be judged. Equally important is *"fidelity to the* *public's values*," Donahue maintains, including how well privatization "deters opportunism and irresponsibility and promotes faithful stewardship."

Judged by this standard, contracting out medical services in Florida's prisons was a flagrant betrayal of the public's trust. Yet a case could be made that it was nothing of the sort—that, to the contrary, Wexford and Corizon delivered exactly what the public expected and perhaps secretly desired. Treating prisoners humanely was not what impelled officials in Florida and other states to embrace privatization after all. Saving money was. "It's a decision that's best for the taxpayers," a DOC official stated at the time. The contract Florida awarded to Wexford and Corizon obligated the companies to spend 7 percent less than the state had. How could this obligation be met without endangering lives? Few people bothered to ask. To deter profiteering, John Donahue argues that private companies entrusted to perform public functions should be subject to vigorous oversight, particularly for tasks whose outcomes are hard to measure. Florida adopted a laxer approach. One year before privatizing prison care, it gutted the Correctional Medical Authority, an independent agency created in 1986 to monitor the quality of medical care that prisoners received.

To be sure, the Wexford and Corizon contract did not explicitly state that profiteering at the expense of vulnerable patients would be tolerated. But as Everett Hughes noted in his essay on dirty work, when it came to the treatment of certain "out-groups," spelling such things out wasn't necessary, particularly when distance and social isolation could achieve the same result. "The greater their social distance from us," Hughes observed of these out-groups, "the more we leave in the hands of others a sort of mandate by default to deal with them on our behalf." The mandate by default given to Wexford and Corizon was to provide grossly substandard care at minimum cost. For the companies' CEOs and shareholders, this arrangement had its advantages, boosting earnings and revenue. For the on-site staff left to do the dirty work of arranging "courtesy drops" for people like Kelly Green, it was another matter.

As it happens, neither Wexford nor Corizon ended up doing this work in Florida for long. Pat Beall's exposé in *The Palm Beach Post* finally prompted some lawmakers to call for greater oversight. In 2015, after state inspectors issued a scathing report about the quality of care delivered by Corizon at a women's prison in Ocala, where patients were cut off from psychiatric medications and improperly placed in isolation cells, Corizon opted not to renew its contract with the state. Two years later, after investigators found "life threatening" deficiencies at another prison where Wexford was in charge, the DOC announced that it was ending its contract with the firm. The glaring deficiencies had persuaded officials in Florida to adopt a new approach, it appeared. But the new approach turned out to be a familiar one—a no-bid contract that was soon awarded to another private company, Centurion of Florida, that was viewed more charitably in Tallahassee, perhaps because of the charity it disbursed to state lawmakers. Before receiving the contract, Centurion gave $765,000 to the Florida Republican Party and another $160,000 to "Let's Get to Work," the political committee of Rick Scott, who, in 2018, ran for Senate and won.

"NOWHERE TO GO"

There were, of course, other ways that Florida could have lowered health expenditures in its prisons, most notably by reducing the number of mentally ill people entering them in the first place. Back in 2008, a special task force convened by a group of advocates and judges issued a report concluding that jails and prisons had become "the unfortunate and undeserving safety nets" for a vast pool of mentally ill people who shuffled back and forth between the correctional system and the streets. Every year, the task force found, "as many as 125,000 people with mental illnesses requiring immediate treatment are arrested and booked into Florida's jails. The vast majority of these individuals are charged with minor misdemeanor and low-level felony offenses that are a direct result of their psychiatric illnesses."

According to the task force, this was not only inhumane, causing "homelessness, increased police injuries, increased police shootings of people with mental illnesses." It was also expensive, with money wasted on costly "back-end" services—overburdened courts, overcrowded prisons—that might have been saved if fewer people fell through the cracks of the state's patchwork mental health system. Florida spent a quarter-billion dollars annually to restore mentally ill people to competency simply so they could stand trial, for example. Meanwhile, half of all adults and one-third of all children with serious mental illnesses had "no access" to psychiatric care in the community.

The chair of the task force that published these findings was Judge Steve Leifman of Florida's Eleventh Judicial Circuit, which spanned Miami-Dade County. When Leifman was seventeen, he interned for a Florida politician who received a letter about a teenager languishing at a state psychiatric hospital. Leifman decided to visit the hospital, where he found the teenager shackled to a bed. Before leaving, he saw naked men hosed down by a guard as though they were zoo animals. Decades later, people with severe mental illnesses were spared such indignities. They were showing up in Leifman's courtroom instead, where they faced the grim prospect of either going to jail or returning to the streets. Dismayed by these bleak alternatives, Leifman pressed local officials to launch a project that would divert mentally ill people who posed no threat to public safety into treatment instead. It took him six years to secure funding for the program, which was soon hailed as a success, boasting a recidivism rate of just 6 percent for individuals who completed it. Over the course of a decade, several thousand people in Miami-Dade County were diverted out of the criminal justice system. Leifman launched another program to train police officers to respond more appropriately to people in psychiatric crises.

Similar reforms have been adopted in other cities, among them San Antonio, where, in 2008, a mental health crisis center founded by a group of local judges and police officers began diverting mentally ill people out of the criminal justice system. Within a decade, it had treated fifty thousand people and saved taxpayers an estimated fifty million dollars. But as admirable as these efforts were, their reach was limited in a society that chose to allow so many people with severe mental illnesses to go without care, a problem that had grown only more severe over time. After the 2008 financial crisis, states slashed five *billion* dollars in mental health services from their budgets, noted a report in *USA Today*, leaving vast numbers of poor mentally ill people with "nowhere to go." Where many eventually did go was onto the caseloads of people like James Fisher, an assistant public defender I met one day in Orlando. Every day, Fisher told me, he watched mentally ill people arrested for "quality of life" offenses cycle through a system that was woefully ill-equipped to deal with them.

Nearly all of the offenders came from poor neighborhoods where afford-able mental health services did not exist. "The population we deal with doesn't have resources to call on anyone but the police," he said.

Some members of this population avoided jail by sleeping on the streets. Others wound up in places like the Lowell Correctional Institu-tion, a prison in central Florida where, on August 21, 2019, a bipolar woman named Cheryl Weimar complained that she was in too much pain to bend down and clean toilets during a work-duty shift because of a hip injury. A group of guards dragged her outside, to an area without security cameras, and beat her so severely that some witnesses thought they were whacking a dead body. Though Weimar survived, the assault left her paralyzed, and showed how little had changed in Florida's prisons since the abuses at Dade had made headlines. In the seven years since Darren Rainey's death, the system remained chronically underfunded, and use-of-force incidents had risen more than 50 percent.

In some other states, the situation was less dire, thanks to advocates who had succeeded in pushing reforms—less reliance on solitary con-finement, more independent oversight—to improve conditions in pris-ons. Yet in the view of Steve J. Martin, a federal monitor who had spent decades investigating the use of force in prisons for various government agencies, Florida was not an anomaly. In America's prisons, "institutional brutality is deeply ingrained and persistent," wrote Martin in July 2020, a few weeks after the asphyxiation of George Floyd by a white police officer in Minneapolis sparked nationwide protests against racism and police brutality. The violence in prisons was more hidden, Martin noted, not least because most correctional systems did not gather or publish data on it, but his experience had led him to believe that the victims looked a lot like the people who were disproportionately subjected to excessive force by the police. In the cases that Martin had investigated, "the pris-oners who died were disproportionately black," he wrote. "Many also had mental impairments."

In our last face-to-face conversation, at a café not far from her home, Harriet Krzykowski told me that since she'd stopped working at Dade, she'd done a lot of reading about solitary confinement and the forces that had turned jails and prisons into America's largest mental health institutions. The reading had helped her put her experience in context. She now understood that the abuses she'd witnessed were part of a much bigger story, she said, in a society that treated the mentally ill as "throwaway people." Knowing this alleviated some of the guilt she felt about what had happened to Darren Rainey.

Before we parted ways, I mentioned that on my next visit to Florida, I was planning to reach out to Rainey's family. Was there anything she wanted me to tell them? I asked her. She fell silent.

"I'm sorry," she said finally, her eyes brimming with tears. "It should never have happened. The conditions that allowed that to happen should never have been in place, *ever*, for anybody."

Some time later, I pulled up to a one-story brick bungalow on West La Salle Street, in downtown Tampa. The house had metal bars on the windows and a moldering gray facade. It sat at the end of a narrow block lined with row houses, in one of Tampa's poorest neighborhoods. Waiting by the entrance was its owner, Andre Chapman, who directed me into a small living room dominated by a large couch.

Andre Chapman was Darren Rainey's brother. He was in his midfifties, a tall, barrel-chested man with gentle brown eyes and an air of quiet forbearance. He'd agreed to meet me after I had called him on the phone and told him that I wanted to know more about his brother's life. After welcoming me inside, Andre started filling in the details, beginning with their childhood, when the members of his family used to fill the pews of the Baptist church across the street on Sundays. Often, he and Darren would be flanking their father, Grady, who was originally from Georgia and had moved to Tampa to be closer to his mother. At the time, the neighborhood was home to a mix of working-class and professional Black families, Andre said. It had since grown more run-down, with an outsize number of vagrants and mentally disturbed people roaming some of the seedier blocks. "I'm talking about some serious mental people," Andre said.

Nothing in Darren's childhood marked him as destined for such a fate, Andre told me. To the contrary, his brother was an honors student who breezed through grammar school and later, under the tutelage of an uncle named Bo, became one of the neighborhood's more proficient checkers players and gamblers. "Oh, he whupped everybody on the street," Andre recalled with a smile. Handsome and outgoing, Darren was also a "ladies' man" who rarely lacked for female company, Andre said, which turned out to be a mixed blessing. In Andre's view, the turning point in his brother's life came in his early twenties, when he fell in love with a woman who was in the military. One night, after they'd been drinking, she pulled out a gun and fired it at Darren's chest. The bullet pierced a lung, but the woman pleaded with Darren not to press charges and he agreed, thinking it would prove his devotion and consummate their love. Soon thereafter, she disappeared from his life.

"That was his downfall," Andre said. "I think it just shook his world a little bit." Yet if being shot in the chest by his lover shook Darren, it did not rob him of his decency, Andre maintained. To give me a sense of his brother's character, Andre invited a neighbor to join us, an elderly woman who rented out the back unit of the gray house. Darren used to come by her place to rake the leaves and do other odd jobs, she said after taking a seat on the couch. In exchange, she would pay him a modest sum or fix him something to eat—cabbage, corn bread, or peanut butter and jelly sandwiches, for which he had a particular fondness. "We just ended up getting close," she said, to the point that she considered Darren family, part of a close-knit circle she called "the La Salle Street posse." In recent years, the posse's ranks had thinned as older friends of hers had begun passing away, which made her miss his presence all the more. "Darren was my handyman," she said. "Darren came to see me every day. I don't care where I was, when I got off of work Darren be sitting on the porch."

In the period before he was arrested and sent to Dade, Andre told me that Darren had been living with him in the same house we were sitting in. The arrest was for cocaine possession. Yet Andre insisted that although Darren smoked cigarettes, he was not doing coke at the time—or, for that matter, at *any* time. "He ain't ever do no cocaine, no weed, nothing like that," he said. Andre was even more emphatic about another point, which is that his brother was not schizophrenic, as had been reported in the *Miami Herald*. "That was a label that they put on Darren when he went to prison," he said.

I heard a different view from Peter Sleasman, an attorney with Disability

Rights Florida, the organization that sued the Florida DOC for subjecting mentally ill incarcerated people at Dade to systemic abuse. While nothing in Darren's records indicated he had ever committed a violent crime, he had numerous convictions for drug offenses, Sleasman told me. The files also contained evidence of persistent mental health problems. "There were a significant number of cases where he was found incompetent to stand trial and was sent to a state hospital for long durations," he said. In Sleasman's view, the records only underscored why prison was "the last place Mr. Rainey belonged." Like so many mentally ill Floridians, he said, Darren would have been far better off getting "drug treatment or community mental health treatment."

In addition to learning more about Darren, I wanted to know how his family found out about what happened at Dade and whom they held responsible. When Darren died, Andre told me, a prison chaplain called him to express his condolences but made no mention of abuse. Andre also spoke to a detective who was no more forthcoming. "I thought he just collapsed in the shower," Andre said. "I'm thinking he had a heart attack." No one from the Miami-Dade County medical examiner's office bothered to send the family Darren's autopsy photos, Andre said. No one from Corizon reached out. Because he suspected no wrongdoing, Andre agreed to have Darren cremated without going to Miami to examine his body beforehand. A few months later, he got a call from Peter Sleasman, who told him that his agency suspected his brother had been scalded to death. Andre was stunned. A soft-spoken man with a calm bearing, he was not given to indignation or sweeping judgments. Yet as he recounted how little he'd been told, he could not disguise his rage. "Y'all took him like some type of rat to be experimented on, and you didn't have the heart to tell me how he died—what really happened?" he said, shaking his head. The autopsy leaked to the press, which described the death as an accident, "was a slap in the face," he added.

The fact that no one was held accountable compounded the insult. "They did the crime; they need to do the time," Andre said. Did "they" include the mental health staff at Dade? I asked him. Andre paused. Then he shook his head. The counselors were just earning a paycheck to support their families, he said. He didn't blame them. He was less forgiving of the guards directly involved, and of other, more powerful actors. After consulting a lawyer, Andre and his family filed a civil rights lawsuit against Florida, Corizon, and some of the guards and prison officials who had worked at Dade. The suit eventually led to a settlement, in which the defendants agreed to pay $4.5 million in damages. The agreement was not exactly an admission of responsibility. To the contrary, Martha Harbin, director of external relations for Corizon, cited it as proof of her company's generosity and blamelessness, telling the *Miami Herald*, "Although every defense lawyer in the case knew Corizon had no liability for Mr. Rainey's death, Corizon contributed $100,000 to the settlement in order to bring the case to a conclusion." Even so, for the Rainey family, the settlement offered a measure of closure—those members of the family who were still around, that is. Andre had since moved to South Carolina. He'd come to Tampa to visit his oldest daughter, Lekesia, who was living in the gray house on West La Salle Street with her young daughter. Returning to the neighborhood brought back "a lot of memories," he said. Among them was a conversation he'd had with his father, Grady, shortly before he passed away, in 2007. Grady's death had devastated Darren, who was extremely close to their father. "That was his backbone," Andre said. Before he died, Andre told me, his father pulled him aside and asked him to look after Darren. "My dad's words were to take care of him, watch out for him," he recalled.

Before I left, Andre reached into a closet and pulled out a shoebox filled with family photographs. He reached into the pile of photos, fished one out, and smiled. It was a picture of his father in a white T-shirt and denim overalls, his usual attire, marking him as a transplant from rural Georgia ("Georgia boy" was his nickname, Andre said). Andre pulled out another photo, a framed portrait of a handsome young man with a stylish Afro and a winsome smile. He was standing next to an attractive woman in a strapless blouse. It was a picture of Darren and his ex-wife, Andre said. The woman in the picture was also the mother of Darren's daughter, he told me, a daughter who was no longer alive. She died of a heart attack at the age of sixteen, not long after learning the news about her father, the last in a string of deaths in the family that had begun with Grady's.

Andre didn't tell me much about Darren's relationship with his daughter. He did say that they were now together, after he had arranged to have Darren's cremated remains placed inside the casket in which she was buried. Andre stared at the picture for a while. "I've been trying to"—his voice caught—"just come to grips." Then he put the picture down, rubbed his reddened eyes, and walked me to my car outside.

PART II

BEHIND THE SCREENS

4

Joystick Warriors

In the spring of 2006, Christopher Aaron started working twelve-hour shifts in a windowless room at the Counterterrorism Airborne Analysis Center in Langley, Virginia. He sat before a wall of flat-screen monitors that beamed live, classified video feeds from drones hovering over distant war zones. On some days, Chris discovered, little of interest appeared on the screens, either because a blanket of clouds obscured visibility or because what was visible—goats grazing on an Afghan hillside, families preparing meals—was mundane, even serene. Other times, what unspooled before his eyes was jarringly intimate: coffins being carried through the streets after drone strikes; a man squatting in a field to defecate after a meal (the excrement generated a heat signature that glowed on infrared); an imam speaking to a group of fifteen young boys in the courtyard of his madrassa. If a Hellfire missile killed the target, it occurred to Chris as he stared at the screen, everything the imam might have told his pupils about America's war with their faith would be confirmed.

The infrared sensors and high-resolution cameras affixed to drones made it possible to pick up such details from an office in Virginia. But as Chris learned, identifying who was in the crosshairs of a potential drone strike wasn't always so easy. The feed on the monitors could be grainy and pixelated, making it easy to mistake a civilian trudging down a road with a walking stick for an insurgent carrying a weapon. The figures on-screen often looked less like people than like faceless gray blobs. How certain could Chris be of who they were? "On good days, when a host of environmental, human, and technological factors came together, we had a strong sense that who we were looking at was the person we were looking for," he said. "On bad days, we were literally guessing."

Initially, the good days outnumbered the bad ones for Chris. He wasn't bothered by the long shifts, the high-pressure decisions, or the strangeness of being able to stalk—and potentially kill—targets from thousands of miles away. Although Chris and his peers spent more time doing surveillance and reconnaissance than coordinating strikes, sometimes they would relay information to a commander about what they saw on-screen, and "sixty seconds later, depending on what we would report, you would either see a missile fired or not," he said. Other times, they would trail targets for months. The first few times he saw a Predator drone unleash its lethal payload—the camera zooming in, the laser locking on, a plume of smoke rising above the scorched terrain where the missile struck—he found it surreal, he told me. But he also found it awe inspiring, experiencing a surge of adrenaline and exchanging congratulatory high fives with the other analysts in the room.

Chris's path to the drone program was unusual. He grew up in Lexington, Massachusetts, in a home where red meat and violent video games were banned. His parents were former hippies who marched against the Vietnam War in the 1960s. But Chris revered his grandfather, a quiet, unflappable man who served in World War II. He also had a taste for exploration and for tests of physical fortitude: hiking and wandering through the woods in Maine, where his family vacationed every summer; wrestling, a sport whose demand for martial discipline captivated him. Chris attended the College of William & Mary in Virginia, where he majored in history. A gifted athlete, he cut a charismatic figure on campus, cultivating an air of independence and adventurousness. One summer, he traveled to Alaska alone to work as a deckhand on a fishing boat.

During his junior year, in 2001, Chris woke up one morning to a phone call from his father, who told him that the Twin Towers and the Pentagon were on fire. Chris thought instantly of his grandfather, who had served for three years as a military police officer on the European front after the attack on Pearl Harbor. He wanted to do something similarly heroic. A year later, after spotting a pamphlet at the William & Mary career-services office for the National Geospatial-Intelligence Agency, a national-security agency that specializes in geographic and imagery analysis, he applied for a job there.

Chris began working as an imagery analyst at the NGA in 2005, studying satellite pictures of countries that had no link to the "war on terror." Not long after he arrived, an email circulated about a Department of Defense task force that was being created to determine how drones could help defeat al-Qaeda. Chris answered the call for volunteers and was soon working at the Counterterrorism Airborne Analysis Center. He found it exhilarating to participate directly in a war he saw as his generation's defining challenge. His pride deepened as it became clear that the task force was having a significant impact and that the use of drones was increasing.

Chris spent a little over a year at the task force, including several months in Afghanistan, where he served as the point of contact between the drone center in Langley and Special Forces on the ground. After this, he worked for a private military contractor for a while. In 2010, an offer came from another contractor involved in the drone program to serve as an imagery-and-intelligence analyst. But as Chris mulled the terms, something strange happened: he began to fall apart physically. The distress began with headaches, night chills, joint pain, a litany of flu-like symptoms that, every few weeks, would recur. Soon, more debilitating symptoms emerged: waves of nausea, eruptions of skin welts, chronic digestive problems. Chris had always prided himself on his physical fitness. Now, suddenly, he felt frail and weak, to the point that working for the contractor was out of the question. "I could not sign the paperwork," he said. Every time he sat down to try, "my hands stopped working—I was feverish, sick, nauseous."

Chris went back to Lexington to live with his parents and try to recuperate. He was twenty-nine years old and in the throes of a breakdown. "I was very, very unwell," he said. He consulted several doctors, none of whom could specify a diagnosis. In desperation, he experimented with fasting, yoga, Chinese herbal medicine. Eventually, his health improved, but his mood continued to spiral. Chris couldn't muster any motivation. He spent his days enveloped in a fog of gloom. At night, he dreamed that he could see—up close, in real time—innocent people being maimed and killed, their bodies dismembered, their faces contorted in agony. In one recurring dream, he was forced to sit in a chair and watch the violence unfurl. If he tried to avert his gaze, his head would be jerked back in place so that he had to continue looking.

"It was as though my brain was telling me: Here are the details that you missed out on," he said. "Now watch them when you're dreaming."

CONSEQUENCES

A few years before Christopher Aaron entered the drone program, a U.S. Army veteran named Eric Fair applied for a job to work as an interrogator in Iraq. Fair was assigned to Abu Ghraib, a prison on the outskirts of Baghdad where American occupation forces began housing detainees after the invasion of Iraq. In April 2004, a cache of photos leaked to the press revealed the sadistic abuse to which many of these detainees were subjected, showing naked prisoners stacked in pyramids, hung upside down, and dragged around on dog leashes as U.S. soldiers smiled and flashed the thumbs-up sign. The photos shocked many Americans and embarrassed President George W. Bush, who had insisted "we do not torture" even as the pictures contradicted this claim. They did not shock Fair, who was not assigned to work in the cellblock at Abu Ghraib where the worst abuses took place but who meted out plenty of harsh punishment of his own. During one interrogation, Fair slammed a prisoner into a wall. He put detainees into "stress positions" and witnessed the use of devices such as the Palestinian chair, onto which detainees were bound with their weight thrust forward and their hands behind their back. A Presbyterian, Fair later published a memoir, *Consequence*, in which he described the crisis of faith and recurrent nightmares that he experienced when he returned home. "I am a torturer," he wrote. "I have not turned a corner or found my way back. I have not been redeemed. I do not believe that I ever will be."

Torturers like Fair dirtied their hands in ways that were visceral and tactile, which, in turn, made many Americans feel dirtied. The pilots and sensor operators in the drone program, by contrast, carried out "precision" strikes on video screens, an activity that seemed a lot cleaner. As the journalist Mark Mazzetti noted in his book, *The Way of the Knife*, remote-control killing seemed like "the antithesis of the dirty, intimate work of interrogation." The shift away from interrogation and torture was overseen by Bush's successor, Barack Obama, who came into office determined to reduce America's military footprint in other countries and to alter the moral tenor of the "war on terror," an expansive phrase that he quickly disavowed. At the start of his first term, Obama signed an executive order banning the use of torture and calling for the closure of the Guantánamo Bay detention camp, where detainees had been subjected to waterboarding and other so-called enhanced interrogation techniques. During his tenure, Obama also dramatically increased targeted drone assassinations, authorizing roughly five hundred lethal strikes outside active conflict zones, ten times the number under Bush.

Early on, when it was first becoming clear that capturing and interrogating suspected terrorists was giving way to secret "kill lists" and targeted assassinations, even some hawkish observers questioned whether the United States knew whom it was killing and whether citizens understood the gravity of what was being done in their name. "Every drone strike is an execution," Richard Blee, a former CIA officer who headed the unit responsible for hunting down Osama bin Laden, told Mazzetti. "If we are going to hand down death sentences, there ought to be some public accountability and some public discussion about the whole thing." For more than two decades, the United States had barred targeted assassinations, a prohibition codified in an executive order that was signed in 1976, after the Church Committee exposed various "illegal, improper, or unethical" activities carried out by the CIA and other government agencies during the Cold War, including a slew of assassination attempts. The executive order established a norm against targeted killings that held sway until 2001, both because policy makers were aware of the order and because of legal and ethical concerns. But if, as Blee maintained, overturning this norm was a serious matter that merited discussion, you would never have known it from the level of public debate that took place. The escalation of the drone wars was met with strikingly little congressional or popular opposition, both under Obama and under his eventual replacement, Donald Trump. In 2016, Trump campaigned as a critic of expansive foreign interventions, which he promised to end. At the same time, Trump made it clear that under his watch the United States would feel even less constrained to carry out extrajudicial killings than under Obama, striking not only militants but also their families. In Trump's first two years in office, more drone strikes were launched in Yemen, Somalia, and Pakistan (three undeclared war zones) than during Obama's entire presidency, and civilian casualties from airstrikes in Afghanistan surged. U.S. commanders were also given free rein to hit a wider array of targets, among them the Iranian general Qassem Soleimani, a high-ranking state official who, on January 3, 2020, was killed by a missile fired from an MQ-9 Reaper drone. Agnes Callamard, the UN special rapporteur on extrajudicial, summary, or arbitrary executions, described the strike as a violation of international law that could set an alarming precedent. "It is hard to imagine that a similar strike against a Western leader would not be considered as an act of war," she said.

While human rights advocates like Callamard voiced concern, the public fell silent. Unlike the harsh interrogation methods that were adopted after 9/11, which sparked vigorous debate about the morality of torture and indefinite detention, drone warfare scarcely registered in public discourse. One explanation for this is that since the Vietnam era and the end of the draft, Americans had grown increasingly disengaged from the wars fought in their name. If something morally troubling happened on the distant battlefields where the conflicts raged, it was easy enough for the public to ignore. On the rare occasions when civilians did pay closer attention, as with the debate about torture during the Bush era, the outrage and opprobrium tended to be directed at individual perpetrators rather than at the system within which they operated. In the case of Abu Ghraib, for example, the blame fell on low-ranking reservists like Charles Graner, who was sentenced to ten years for his role in abusing Iraqi detainees, and Lynndie England, who was given a three-year sentence and dishonorably discharged. No senior officials were held accountable. As with the brutality in America's prisons, which was often pinned on a handful of sadistic guards, the violence in America's military campaigns could be attributed to a few "bad apples," diverting attention from the system of war, which retained its moral legitimacy.

Another factor deterring robust debate about the drone program was that it was swathed in secrecy. According to U.S. officials, the laser-guided missiles fired from drones caused minimal collateral damage and were authorized for use only against high-level targets who posed an "imminent threat" to national security and could not be captured. Like so much else about the drone program, however, the identities of the dead were not shared with the public. As a 2017 report by the Columbia Law School Human Rights Clinic and the Sana'a Center for Strategic Studies noted, the U.S. government officially acknowledged just 20 percent of more than seven hundred reported drone strikes since 2002 in Pakistan, Somalia, and Yemen. It also failed to respond to nongovernmental organizations that had requested explanations about strikes in which "credible evidence of civilian casualties and potential unlawful killings" had surfaced.

The criteria used to approve drone strikes, the number of civilian deaths: all of this was kept hidden. According to the Bureau of Investigative Journalism, a London-based organization that independently tracked America's drone program, U.S. drone strikes killed between 8,858 and 16,901 people outside acknowledged war zones between the date it began collecting data and the end of 2020, including as many as 2,200 civilians. But because most of these strikes occurred in remote areas that were inaccessible to reporters, the public rarely heard about them. The sanitized language that public officials used to describe drone strikes ("pinpoint," "surgical") reinforced the perception that drones had turned war into a bloodless exercise.

The lack of transparency prevented ordinary citizens from knowing even basic facts about who their country was bombing and why. On the other hand, one could argue that the opaqueness was convenient, sparing citizens from having to think too much about a campaign of endless war to which many were tacitly resigned. As Everett Hughes might have noted, remote-control killing had an "unconscious mandate" from the public, solving a problem in a nation that had grown disillusioned with the "war on terror" but didn't necessarily want real constraints placed on America's use of force. As exhausted by war as Americans had become, many were accustomed to the idea that the United States could exercise its military might on a geographically limitless scale and perhaps saw this as necessary, both to keep America safe and to project U.S. power.

Tuning out the drone campaign was all the easier because no U.S. soldiers risked dying in it. In contrast to messy ground wars like the invasion of Iraq, which cost trillions of dollars and thousands of American lives, drones fostered the alluring prospect that terrorism could be eliminated at the press of a button. It was war without risk—without consequences, at least for our side, which made it easy for "good people" to keep the dirty business of targeted assassinations out of mind.

The "joystick warriors" pressing the buttons also experienced no consequences, some analysts suggested, owing to distance and technology, which sheared war of its moral gravity, turning killing into an activity as carefree as playing a video game. Serving as a drone operator was not, to borrow Erving Goffman's terminology, "people work" that required daily interaction with "human materials," as was the case for prison guards like Bill Curtis and for military interrogators like Eric Fair. It was detached, impersonal desk work, filtered through technology that desensitized the human beings involved to the consequences of their actions. In 2010, Philip Alston, the former UN special rapporteur on extrajudicial, summary, or arbitrary killings, warned that because drone operators "are based thousands of miles away from the battlefield and undertake operations entirely through computer screens and remote audio-feed, there is a risk of developing a 'PlayStation' mentality to killing."

It was a logical theory, albeit one formulated without input from pilots who had actually participated in remote combat operations. When military psychologists began talking to imagery analysts and sensor operators about their experiences, a different picture emerged. In one survey, Wayne Chappelle and Lillian Prince, researchers for the School of Aerospace Medicine at Wright-Patterson Air Force Base, drew on interviews with 141 intelligence analysts and officers involved in remote combat operations to assess their emotional reactions to killing. Far from exhibiting a sense of carefree detachment, three-fourths of the subjects reported feeling negative emotions—grief, sadness, remorse—related to killing. Many experienced these "negative, disruptive emotions" for a long duration (defined as one month or more). Another study conducted by the U.S. Air Force found that drone analysts in the "kill chain" were exposed to more graphic violence—seeing "destroyed homes and villages," witnessing bodies burned alive—than most Special Forces on the ground.

For the people staring at the screens, remote-control killing wasn't so clean, the studies suggested. It was messy and disturbing, albeit in ways that differed from conventional warfare. Because they never set foot on the battlefield, drone operators were not exposed to roadside bombs, a common cause of brain injuries and PTSD among veterans of America's recent wars. They did not experience haunting flashbacks of improvised explosive devices detonating on the streets they were patrolling and destroying their Humvees.

What, then, *did* they experience? One morning, I visited the Creech Air Force Base in Nevada to find out. Located forty miles north of Las Vegas, Creech is a constellation of windswept airstrips surrounded by sagebrush and cactus groves. The base is home to nine hundred drone pilots and sensor operators who fly missions with MQ-9 Reaper drones. It is also home to a group of embedded psychologists, chaplains, and physiologists called the Human Performance Team, which was established to address the rising levels of stress and burnout in the drone program.

All members of the Human Performance Team possess the security clearances required to enter the ground control stations where drone pilots do their work, in part so that they can get a glimpse of what they experience. A psychologist on the team named Richard (who, like most of the airmen I spoke to, asked to be identified only by his first name) told me that two weeks into the job he poked his head into a ground control station just as the crew was "spinning up for a strike." A veteran of the Marine Corps, he felt his adrenaline rise as he watched the screen flash. Then he put the incident out of mind. A few weeks later, he was at his son's band concert; as the national anthem played and he peered up at the Stars and Stripes, the memory came back. "I'm looking up at the flag, but I could see a dead body," he said. He was shaken, but he couldn't say anything to his family, because the operation was classified.

Drone warriors shuttled back and forth across such boundaries every day. When their shifts ended, the airmen and airwomen drove to their subdivisions alone, like clerks in an office park. One minute they were at war; the next they were at church or picking up their kids from school. A retired pilot named Jeff Bright, who served at Creech for five years, described the bewildering nature of the transition. "I'd literally just walked out on dropping bombs on the enemy, and twenty minutes later I'd get a text: Can you pick up some milk on your way home?" he said. Bright enjoyed serving in the drone program and believed that he was making a difference. But other airmen in his unit struggled to cope with the stress, he told me; there were divorces, and some cases of suicide.

Unlike office park employees, drone operators could not reveal much about how their day went because of classification restrictions. Unlike conventional soldiers, they weren't bolstered by the group solidarity forged in combat zones. Richard told me that when he was in the U.S. Marines, "there was a lot of camaraderie, esprit de corps." Although drone operators could get close to their coworkers, at the end of every shift they went home to a society that had grown increasingly disconnected from war. The disconnect was especially profound at Creech, where the desert scrub surrounding the base soon gave way to billboards advertising live entertainment and gambling resorts in Las Vegas, where many of the service members lived. An hour after visiting the base, I wandered the Strip, watching tourists snap selfies in front of various landmarks—the Fountains of Bellagio, the High Roller at the Linq—before making their way to the endless procession of lounges, nightclubs, casinos, and all-you-can-eat buffets. The atmosphere of gaudy excess underscored the strangeness of the fact that forty-five minutes away a war was being fought in the name of these very same revelers.

Before the drone personnel at Creech made their way home, some dropped by the Airman Ministry Center, a low-slung beige building equipped with a foosball table, some massage chairs, and several rooms where pilots and sensor operators could talk with clergy. A chaplain named Zachary told me that what most burdened the airmen he spoke to was not PTSD. It was inner conflicts that weighed on the conscience. Zachary mentioned one pilot he met with who asked, "I'm just curious: What is Jesus going to say to me about all the killing I've done?" Despite their distance from the battlefield, drone operators' constant exposure to "gut-wrenching" things they watched on-screen—sometimes resulting directly from their own split-second decisions, other times from their *inability* to act—could cause them to lose their spiritual bearings and heighten their risk of sustaining a very different kind of battle scar: a wound Zachary described as a "moral injury."

The term was not new. It first appeared in the 1994 book *Achilles in*

Vietnam, by the psychiatrist Jonathan Shay, who drew on Homer's epic war poem, the *Iliad*, to probe the nature of the wounds afflicting veterans of the Vietnam War. Shay read the *Iliad* as "the story of the undoing of Achilles' character," which, he argued, changes when his commander, Agamemnon, betrays his sense of "what's right," triggering disillusionment and the desire "to do things that he himself regarded as bad." Experiencing such disillusionment might not seem as traumatic as being shot at or seeing a comrade die. Shay disagreed. "I shall argue what I've come to strongly believe through my work with Vietnam veterans: that moral injury is an essential part of any combat trauma," he wrote. "Veterans can usually recover from horror, fear, and grief once they return to civilian life, so long as 'what's right' has not also been violated."

After 9/11, the term "moral injury" began to appear more frequently in the literature on the psychic wounds of war, but with a slightly different meaning. Where Shay emphasized the betrayal of what's right by authority figures, a new group of researchers expanded the focus to include the anguish that resulted from "perpetrating, failing to prevent, or bearing witness to acts that transgress deeply held moral beliefs," as a 2009 article in the journal *Clinical Psychology Review* posited. In other words, from wounds sustained when soldiers wading through the fog of war betrayed *themselves*, through harmful acts they perpetrated or watched unfold. This definition took shape against the backdrop of the wars in Iraq and Afghanistan, chaotic conflicts in which it was difficult to distinguish between civilians and insurgents and in which the rules of engagement were fluid and gray.

One author of the *Clinical Psychology Review* article was Shira Maguen, a researcher who began to think about the moral burdens of warfare while counseling veterans at a PTSD clinic in Boston. Like most Veterans Affairs psychologists, Maguen had been trained to focus on the aftershocks of fear-based trauma—IED blasts that ripped through soldiers' Humvees, skirmishes that killed members of their unit. The link between PTSD and such "life-threat" events was firmly established. Yet in many of the cases she observed, the source of distress seemed to lie elsewhere: not in attacks by the enemy that veterans had survived, but in acts they had observed or carried out that crossed their own ethical lines.

Soldiers were not, of course, the only people who risked committing such transgressions. All of the counselors I interviewed at the Dade Correctional Institution struggled with inner conflicts related to horrifying things they'd witnessed but failed to prevent. What kind of person was she? Lovita Richardson wondered after seeing a prisoner bound to a chair get bludgeoned and not intervening to help him. "Why didn't I do more?" Harriet Krzykowski asked herself after learning about the "shower treatment." Many of the prison guards I'd interviewed had alluded to incidents where they'd done things they knew they shouldn't, as when Bill Curtis slammed a man to the ground, nearly fracturing his skull. Moral injuries were an occupational hazard for anyone whose job involved "perpetrating, failing to prevent, or bearing witness to acts that transgress deeply held moral beliefs." For most dirty workers, that is.

Among the veterans she counseled, Maguen grew particularly interested in the emotional toll of killing, which was sanctioned in the military but not when defenseless civilians were involved. "I was hearing about experiences where people killed and they thought they were making the right decision," she told me, "and then they found out there was a family in the car." To find out how heavy the burden of killing was, Maguen began combing through the databases in which veterans of conflicts dating back to the Vietnam War were asked if they had killed someone while in uniform. In some cases, veterans were also asked *whom* they killed— combatants, prisoners of war, civilians. Maguen wanted to see if there might be a relationship between taking another life and debilitating consequences like alcohol abuse, relationship problems, outbursts of violence, PTSD. The results were striking: even when controlling for different experiences in combat, she found, killing was a "significant, independent predictor of multiple mental health symptoms" and of social dysfunction.

Later, when she started directing a mental health clinic at a VA hospital in San Francisco, Maguen convened groups where veterans came together and talked about the killing they had done. In the VA no less than in the military, this was a taboo subject, so much so that clinicians often referred to it euphemistically, if at all. To ease the tension, a scene from a documentary was shown at the beginning of each session in which a veteran said, "Out there, it's either kill or be killed. Nothing can really prepare you for war." Afterward, Maguen would ask the veterans in the room a series of questions about how killing had impacted their lives.

Some reacted angrily. Others fell silent. But many seized the opportunity to talk about experiences they later told Maguen they had never spoken about with anyone, not even their spouses and family members, for fear of being judged.

The veterans in Maguen's groups didn't talk a lot about fear and hyperarousal, emotions linked to PTSD. Mostly, they expressed self-condemnation and guilt. "You feel ashamed of what you did," one said. Others described feeling unworthy of forgiveness and love. The passage of time did little to diminish the depth of these feelings, Maguen found. Geographic distance didn't lessen them much either. Maguen recounted the story of a pilot who was haunted by the bombs he had dropped on victims far below. What troubled him was, in fact, precisely his distance from them—that instead of squaring off against the enemy in a fair fight, he had killed in a way that lacked valor. Obviously not all pilots felt this way. But the story underscored the significance of something Maguen had come to regard as more important than proximity or distance in shaping moral injury—namely, how veterans made sense of what they had done. "How you conceptualize what you did and what happened makes such a big difference," she said. "It makes all the difference."

Unlike PTSD, moral injury was not a medical diagnosis. It was an attempt to capture what could happen to a person's identity and soul in the crucible of war, which is why it struck a chord among veterans who did not feel their wounds could be reduced to a medical disorder. "PTSD as a diagnosis has a tendency to depoliticize a veteran's disquietude and turn it into a mental disorder," observed Tyler Boudreau, a marine officer who served in Iraq and came back haunted by doubts about the war's morality. "What's most useful about the term 'moral injury' is that it takes the problem out of the hands of the mental health profession and the military and attempts to place it where it belongs—in society, in the community, and in the family—precisely where moral questions should be posed and wrangled with. It transforms 'patients' back into citizens and 'diagnoses' into dialogue."

Not everyone welcomed this transformation. The meaning and magnitude of moral injury remained contested. "It is not a term widely accepted by the military or the psychological community," Wayne Chappelle, of the School of Aerospace Medicine at Wright-Patterson Air Force Base, told me, adding that he did not believe it was prevalent among drone opera-tors. This was somewhat surprising, because Chappelle was an author of the study revealing that many drone warriors struggled with unresolved negative emotions after strikes, feeling "conflicted, angry, guilty, regret-ful." Then again, perhaps it wasn't so surprising. The idea that war may be morally injurious is a charged and threatening one to many people in the military. Tellingly, Chappelle described moral injury as "intentionally doing something that you felt was against what you thought was right," like the wanton abuse of prisoners at Abu Ghraib. The definition used by research-ers like Maguen was at once more prosaic and, to the military, potentially more subversive: moral injury was sustained by soldiers in the course of doing exactly what their commanders, and society, asked of them.

"MORAL HAZARD"

By the time I met Christopher Aaron, he had spent several years recuper-ating from his experience in the drone program. We first talked at length in a pub, not far from where he was living at the time. Chris was in his mid-thirties, with thick dark hair pulled back in a ponytail. He had a calm, Zen-like bearing, honed in part through yoga and meditation, but there was a trace of worry in his eyes and a degree of circumspection in his voice, especially when pressed for details about particular missions (he could not talk about anything classified, he told me). At the pub, we spoke for two hours and agreed to continue the conversation over lunch the next day so that he could pace himself. As I was heading to that lunch appointment, my cell phone rang. It was Chris, calling to reschedule. Our meeting the previous day had triggered a flood of anxiety, aggravating the pain in his back during the night.

Some analysts in the drone program sensed immediately that their work left an emotional residue. In Chris's case, the feeling unfolded grad-ually, coinciding with a shift in worldview, as his gung ho support for the "war on terror" gave way to glimmers of doubt. The disillusionment crept up in stages, starting, he realized in retrospect, a few months after he returned from Afghanistan. Although he felt proud of the work he'd done to help establish the drone program, he also started to wonder when the war's objective was going to be achieved. It was around this time that his manager asked him if he wanted to obtain resident CIA employment status and become a career intelligence officer, which required taking a lie-detector test used to screen employees. Chris said yes, but halfway through the test, after losing circulation in his arms and feeling hectored by the questions, he got up and abruptly left. The next day, Chris told his manager that he had reconsidered.

Chris ended up taking a trip to California instead, renting a motorcycle and riding all the way up the coast to Alaska, where he spent a week at a monastery on a small island, sleeping in a wood-framed chapel surrounded by spruce trees. Chris had grown up attending an Eastern Orthodox church, and the experience was faith reaffirming. When he went back to the East Coast, he felt refreshed. But he was also out of money, which led him to apply for a job at the only company he figured would hire him—a military-and-intelligence contractor involved in the drone program and the war in Afghanistan.

By now, Chris's idealism had waned. It receded further when, at the end of 2008, the contractor sent him back to Afghanistan. The first time he was there, in 2006, the "war on terror" seemed to be hastening the defeat of al-Qaeda and the Taliban. Now it seemed to Chris not only that progress had stalled but that things were sliding backward. "We were actually losing control of vast areas of the country," he said, even as the number of drone strikes was "four or five times higher" than before. The escalation under President Obama had begun.

As it happened, Chris had taken a copy of George Orwell's *1984* with him to Afghanistan. He had read the book in high school and, like most people, remembered it as a dystopian novel about a totalitarian police state. This time, what stuck in his mind was the book-within-the-book written by Emmanuel Goldstein, the rumored leader of the resistance, titled *The Theory and Practice of Oligarchical Collectivism*. In the book, Goldstein describes the onset of a "continuous" war waged by "highly trained specialists" on the "vague frontiers" of Oceania—an opaque, low-intensity conflict whose primary purpose was to siphon off resources and perpetuate itself. ("The object of waging a war is always to be in a better position in which to wage another war," Goldstein observes.) Chris had an eerie sense that a perpetual war was exactly what the "war on terror" was becoming.

As his disillusionment deepened, events that Chris dismissed before as unavoidable in any war began to weigh more heavily on him. He recalled days when the feed was "too grainy or choppy" to make out exactly who was struck. He remembered joking with peers that "we sometimes didn't know if we were looking at children on the ground or chickens." He also thought back to the times when he would be asked "to give an assessment of a compound where they had suspicion: there was fill-in-the-blank low-level Taliban commander in a remote region in the country. And we had seen other people come in and out of the same compound over the course of the preceding two or three days. They come and say, 'We're getting ready to drop a bomb on there. Are there any people other than the Taliban commander in this compound?' I'd just say no because they don't want to hear 'I don't know.' And then two days later, when they have the funeral procession in the streets that we could observe with the Predators, you'd see, as opposed to carrying one coffin through the streets, they're carrying three coffins through the streets."

Chris kept his misgivings mainly to himself, but his friends noticed a change in him, among them Chris Mooney, who picked him up at the airport when he returned from Afghanistan in 2009. Mooney and Chris had been friends since college, when Chris had exuded a robust sense of confidence and enthusiasm. "He was magnetic," said Mooney, who recalled taking road trips with Chris and marveling at his energy and assertiveness. At the airport, Mooney could scarcely recognize his friend. His affect was flat, his face a solemn mask. They went to dinner, where, at one point, a patron who overheard them talking came up to Chris to thank him for his service. Chris thanked him back, Mooney said, but in a muted tone. Mooney didn't press him for details, but he knew that something was seriously wrong. "It wasn't the same guy," he said.

The renewed interest in moral injury could be viewed as an effort to revisit ethical issues that had been lurking in our narratives of war all along—and to address sources of trauma that some veterans and military analysts recognized before the "war on terror" even began. In his influential 1995 book, *On Killing*, Dave Grossman, a retired army lieutenant and former professor of psychology at West Point, drew on historical studies and the personal accounts of ex-combatants to argue that the psychological costs of killing were often devastating. The novels and memoirs of veterans were filled with characters haunted by such incidents. In Tim O'Brien's *The Things They Carried*, for example, the narrator confesses that he can't shake the image of the Vietnamese man he killed on a footpath with a grenade—his body splayed, blood glistening on his neck. Later, the narrator indicates that he didn't actually kill the man, but he was there and watched him die, "and my presence was guilt enough." Literature could evoke the inner conflicts that played on a loop in the minds of veterans tormented by their troubled consciences.

In the early 1970s, some psychiatrists listened to soldiers talk about such incidents at "rap groups" organized by Vietnam Veterans Against the War. Until this point, soldiers bearing psychic wounds tended to be dismissed by the military as cowards and malingerers. ("Your nerves, hell— you are just a goddamned coward," General George Patton snapped at a soldier in a hospital during World War II.) The Yale psychiatrist Robert Jay Lifton, who sat in on the VVAW rap groups and wrote about the disfiguring effects of killing and participating in atrocities in his 1973 book, *Home from the War*, helped to recast these veterans as sympathetic figures. The participants in the rap groups were burdened not by cowardice but by the guilt and rage they felt at their involvement in a misbegotten war, Lifton argued, "a filthy and unfathomable war" from which they returned "defiled . . . in the eyes of the very people who sent them as well as in their own." In Lifton's view, moral and political questions were inseparable from Vietnam veterans' psychic wounds, to the point that he believed activism to end the war could lessen their guilt and foster healing.

When PTSD was officially recognized in the *Diagnostic and Statistical Manual of Mental Disorders*, in 1980, many hoped it would lead society to reckon more honestly with the ethical chaos of war. The first definition of PTSD included among the potential symptoms not only survivor's guilt but also guilt "about behavior required for survival," language that addressed acts soldiers perpetrated that went against their own ethical codes. Over time, however, the moral questions that animated reformers like Lifton were reduced to "asterisks in the clinician's handbook," notes the veteran David Morris in his book *The Evil Hours*, as military psychologists shifted attention to brain injuries caused by mortar attacks and roadside bombs. One reason for this may be that focusing on such injuries, and on harmful acts in which veterans were the victims, was less threatening to the military. Another is that it might have been less fraught for VA clinicians, who weren't trained to address veterans' moral pain and who "may unknowingly provide nonverbal messages that various acts of omission or commission in war are too threatening or abhorrent to hear," noted the authors of the 2009 article on moral injury in *Clinical Psychology Review*. Avoiding such conversations was particularly untenable with service members returning from Iraq and Afghanistan, who were enmeshed in counterinsurgency campaigns that often involved close-range killing and noncombatants. According to Brett Litz, a clinical psychologist at Boston University and an author of the 2009 article, "35 percent of the traumatic events that led soldiers to seek treatment for PTSD in a recent study were morally injurious events."

When the drone program was created, it seemed to promise to spare soldiers from the intensity (and the danger) of close-range combat. But fighting at a remove could be unsettling in other ways. In conventional wars, soldiers fired at an enemy who had the capacity to fire back at them. They killed by putting their own lives at risk. What happened when the risks were entirely one-sided—when the ethic of survival that prevailed on the battlefield ("it's either kill or be killed") didn't actually apply? In his book *Killing Without Heart*, M. Shane Riza, a retired U.S. Air Force instructor and command pilot, cited a dictum attributed to the French philosopher Albert Camus: "You can't kill unless you are prepared to die." By making them impervious to death and injury, unmanned aerial vehicles turned warriors into assassins, Riza averred, a form of warfare that was bereft of honor. Lawrence Wilkerson, a retired army colonel and former chief of staff to Colin Powell, shared this view, fearing that remote warfare eroded "the warrior ethic," which held that combatants must assume some measure of reciprocal risk. "If you give the warrior, on one side or the other, complete immunity, and let him go on killing, he's a murderer," he said. "Because you're killing people not only that you're not necessarily sure are trying to kill you—you're killing them with absolute impunity."

Unlike conventional soldiers, drone operators were not eligible for Bronze Stars or combat pins adorned with the letter *V*, which stood for "valor." In 2013, the then defense secretary, Leon Panetta, announced that a special "Distinguished Warfare Medal" would be awarded to remote combat operators who had made important contributions to national defense. The proposal elicited a flood of complaints from military veterans, with some dismissing the decoration as a "Nintendo" medal. In the face of this backlash, the military shelved the plan, eventually agreeing to award cyber warriors pins with the letter *R*, which stood for "remote." The negative reaction underscored how, within the military as in society at large, the "joystick warriors" in the drone program were seen as less honorable and courageous than real soldiers who stepped on the battlefield and put their lives on the line. It also underscored an irony, which is that this inferior status derived from the very thing that made drone warfare appealing to politicians and the public—namely, the fact that it enabled America to carry out lethal operations in other countries with no risk of incurring more casualties. Drone operators who killed from afar were very much the agents of a society that, after the protracted conflicts in Iraq and Afghanistan, which squandered hundreds of billions of dollars and thousands of lives, wanted to have the military conduct its business at a minimal cost in blood and treasure, at least for our side.

Langley Air Force Base in Virginia is home to a division of the 480th Intelligence, Surveillance, and Reconnaissance Wing, a unit of six thousand "deployed in place" cyber warriors. They work on what is known as the ops floor, a dimly lit room equipped with a riot of computer screens streaming footage from drones circling over numerous battlefields. Many of the enlistees arrayed around the screens are in their twenties; were it not for their military boots and combat fatigues, they might pass for stock traders or Google employees. But the decisions they make have far weightier consequences. According to a survey by a team of embedded air force researchers who surveyed personnel at three different bases, nearly one in five ISR analysts said they "felt directly responsible for the death of an enemy combatant" on more than ten occasions. One analyst told the researchers, "Some of us have seen, read, listened to extremely graphic events hundreds and thousands of times."

"Overall, I.S.R. personnel reported pride in their mission, particularly supporting successful protection of U.S. and coalition forces," the survey found. But many also struggled with symptoms of distress—emotional numbness, difficulty relating to family and friends, trouble sleeping, and "intrusive memories of mission-related events."

As at Creech, steps had been taken to try to mitigate the stress: shorter shifts, softer lighting, embedded chaplains and psychologists. But the workload of ISR analysts had also increased as drones assumed an increasingly central role in the battle against the Islamic State and other foes. According to Lieutenant Colonel Cameron Thurman, who was the unit's surgeon general during the time I visited, the number of acknowledged missile strikes ordered by Central Command in the United States rose substantially between 2013 and mid-2017, even as the size of the workforce remained unchanged. "You've got the same number of airmen doing the same number of mission hours but with a one thousand percent increase in those life-and-death decisions, so of course their job is going to get significantly more difficult," he said. "You're going to have more moral overload."

A bald man with a blunt manner, Thurman sat across from me in a windowless conference room whose walls were adorned with posters of squadrons engaged in remote combat operations. Also in the room was Alan Ogle, a psychologist who was an author of the survey of the 480th Wing. On the PTSD scale, Ogle said, members of the unit "didn't score high," owing to the fact that few had been exposed to roadside bombs and other so-called life-threat events. What seemed to plague them more, he told me, were some forms of "moral injury."

Two members of the ISR Wing described to me how their work had changed them. Steven, who had a boyish face and sensitive eyes, was originally from a small town in the South and joined the military straight out of high school. Four years later, he told me, he no longer reacted emotionally to news of death, even after the recent passing of his grandmother, with whom he was extremely close. The constant exposure to killing had numbed him. "You're seeing more death than you are normal things in life," he said. On the ops floor, he'd watched countless atrocities committed by ISIS. During one mission, he was surveilling a compound on a high-visibility day when ten men in orange jumpsuits were marched outside, lined up, and, one by one, beheaded. "I saw blood," he said. "I could see heads roll." Ultimately, though, what troubled him most was not bearing witness to vicious acts committed by enemy forces, but decisions *he* had made that had fatal consequences. Even if the target was a terrorist, "it's still weird taking another life," he said. Distance did not lessen this feeling. "Distance brings it through a screen," he said, "but it's still happening, and it's happening because of you."

Another former drone operator told me that screens could paradoxically magnify a sense of closeness to the target. In an unpublished paper that he shared with me, he called this phenomenon "cognitive combat intimacy," a relational attachment forged through close observation of violent events in high resolution. In one passage, he described a scenario in which an operator executed a strike that killed a "terrorist facilitator" while sparing his child. Afterward, "the child walked back to the pieces of his father and began to place the pieces back into human shape," to the horror of the operator. Over time, the technology of drones had improved, which, in theory, made executing such strikes easier but also made what remote warriors saw more vivid and intense. The more they watched targets go about their daily lives—getting dressed, eating breakfast, playing with their kids—the greater the operators' "risk of moral injury," his paper concluded.

This theory was echoed by Shira Maguen's findings. In one study, she discovered that Vietnam veterans who killed prisoners of war had especially high rates of trauma. Maguen believed the reason was that the victims were not strangers to them. "When someone is a prisoner of war, you get to know them," she explained, "you have a relationship with them. You are watching them; you are talking to them. It may be that with drone operators they also know their subjects fairly well: they have watched them, so there's a different kind of relationship, an intimacy."

For Christopher Aaron, the hardest thing to come to terms with was that a part of him had enjoyed wielding this awesome power—that he'd found it, on some level, exciting. In the years that followed, as his mood darkened, he withdrew, sinking into a prolonged funk shadowed by shame and grief. He avoided seeing friends. He had no interest in intimate relationships. He struggled with quasi-suicidal thoughts, he told me, and with facing the depth and gravity of his wounds, a reckoning that began in earnest only in 2013, when he made his way to the Omega Institute in Rhinebeck, New York, to attend a veterans' retreat run by a former machine gunner in Vietnam.

The weather during the retreat was rainy and overcast, matching Chris's somber mood. The discussion groups Chris sat in on, where veterans cried openly as they talked about their struggles, were no more uplifting. But for the first time since leaving the drone program, Chris felt that he didn't have to hide his true feelings. Every morning, he and the other veterans would begin the day by meditating together. At lunch, they ate side by side, a practice called "holding space." In the evenings, Chris drifted into a deep slumber, unperturbed by dreams. It was the most peaceful sleep he'd had in years.

At the Omega Institute, Chris struck up a friendship with a Vietnam veteran from Minnesota, whom he later invited up to Maine. In the fall of 2015, at his friend's suggestion, he went to a meeting at the Boston chapter of Veterans for Peace. Soon thereafter, he began to talk—first with members of the group, later at some interfaith meetings organized by peace activists—about funeral processions he'd witnessed after drone strikes where more coffins appeared than he expected. It was painful to dredge up these memories; sometimes his back would seize up. But it was also therapeutic, a form of social engagement that connected him to a larger community.

At one interfaith meeting, Chris mentioned that he and his colleagues used to wonder if they were playing a game of "whack-a-mole," killing one terrorist, only to see another pop up in his place. He had come to see the drone program as an endless war whose short-term "successes" only sowed more hatred in the long term while siphoning resources to military contractors that profited from its perpetuation. On other occasions, Chris spoke about the "diffusion of responsibility," the whirl of agencies and decision makers in the drone program that made it difficult to know what any single actor had done. This was precisely the way the military wanted it, he suspected, enabling targeted killing operations to proceed without anyone feeling personally responsible. And yet, if anything, Chris felt an *excess* of remorse and culpability, convinced that targeted killings had very likely made things worse.

The relief Chris drew from talking about his experience would not have surprised Peter Yeomans, a clinical psychologist who trained with

Shira Maguen and who ran an experimental treatment for moral injury that was rooted in the sharing of testimonials, initially at weekly meetings where veterans came together to talk among themselves. After ten weeks, the treatment culminated in a public ceremony that the participants invited members of the community to attend. One goal of the treatment was to help veterans unburden themselves of shame, Yeomans told me. Another was to turn them into moral agents who could deliver the truth about war to their fellow citizens and, in turn, broaden the circle of responsibility for their conduct.

One evening, I attended a ceremony in a small chapel on the third floor of the VA Medical Center in Philadelphia, where Yeomans worked. Seated on a stage in the chapel were a number of veterans, among them a slightly built man with an unkempt brown beard who sat with his eyes closed and his hands folded in his lap. His name was Andy, and when invited to speak, he told the audience that he grew up in a violent home where he watched his older brother and baby sister endure abuse, which made him want to "protect the defenseless." After high school, he enlisted in the military and became an intelligence operative in Iraq. One night, on a mission near Samarra, a city in the "Sunni triangle," a burst of sustained gunfire erupted from the second-story window of a house. Andy said he "called air" to deliver a strike. When the smoke cleared from the leveled home, there was no clear target inside. "I see instead the wasted bodies of nineteen men, eight women, nine children," Andy said, choking back tears. "Bakers and merchants, big brothers and baby sisters.

"I relive this memory almost every day," he went on. "I confess to you this reality in the hope of redemption, that we might all wince and marvel at the true cost of war."

The room fell silent as Andy went back to his chair, sobbing. Then Chris Antal, a Unitarian Universalist minister who ran the weekly meetings with Yeomans, invited members of the audience to form a circle around the veterans who had spoken and deliver a message of reconciliation to them. Several dozen people came forward and linked arms. "We sent you into harm's way," began the message that Antal recited and that the civilians encircling the veterans repeated. "We put you into situations where atrocities were possible. We share responsibility with you: for all that you have seen; for all that you have done; for all that you have failed to do," they said. Later, members of the audience were invited to come forward again, this time to take and carry candles that the veterans had placed on silver trays when the ceremony began. Andy's tray had thirty-six candles on it, one for each person killed in the airstrike that he called in.

Yeomans and Antal told me over dinner afterward that they believed audience participation in the ceremony was crucial. Moral injury, they suggested, was exacerbated by society's growing disengagement from war, which left veterans like Andy to struggle with the costs and consequences on their own. Antal added that, in his opinion, grappling with moral injury required reckoning with how America's military campaigns had harmed not only soldiers but also Iraqis and civilians in other countries.

For Antal, broadening the scope to include these civilians was both a spiritual mission and a personal one, because he bore a moral injury of his own. It was sustained when he was serving as an army chaplain in Afghanistan. While there, he attended ceremonies in which the coffins of fallen U.S. soldiers were loaded onto transport planes to be sent home. During one such ceremony, held at Kandahar Airfield, Antal noticed drones taking off and landing in the distance and felt the flicker of conscience. The contrast between the dignity of the ceremony, during which the fallen soldier's name was solemnly announced as taps was played, and the secrecy of the drone campaign, whose victims were anonymous, jarred him. "I felt something break," he told me. In April 2016, Antal resigned his commission as a military officer, explaining in a letter to President Obama that he could not support a policy of "unaccountable killing" that granted the executive branch the right to "kill anyone, anywhere on earth, at any time, for secret reasons." In a doctoral dissertation he later submitted to the Hartford Seminary, Antal reflected on the "moral hazard" created by the covert drone program and the growing reliance on special operations forces, which enabled civilians to "know less, risk less, and thus care less" about "the violence inflicted in their name." Consequently, he wrote, "veterans are often the ones left holding the pain society would rather ignore or forget. Meanwhile, the US military has a presence in almost every country on earth, and has more funding and greater kill capacity than ever before."

The secrecy of the drone program made it all the more essential that the public heard more from service members about what they saw and did.

But it also made it riskier for people who had served in the program to share their stories. Jesselyn Radack, a lawyer for national-security whistleblowers, told me that several former drone operators she represented had suffered retaliation for talking about their experiences (she said one client had his house raided by the FBI and was placed under criminal investigation after speaking on camera with a filmmaker). When Christopher Aaron began speaking publicly about his own past, he contacted Radack, fearing the same thing might happen to him. Initially, nothing did happen. But in June 2017, after someone hacked into his email, a stream of anonymous threats began flooding his inbox. The hostile messages, calling him "scum" and warning him to "shut his big blabbermouth," were also sent to his father, whose email was likewise hacked. The barrage of threats eventually prompted Chris to hire a lawyer to try to identify who was behind the harassment (the attorney he worked with, Joe Meadows of Bean Kinney & Korman, specialized in internet defamation), and to contact both the FBI and the police.

The experience left Chris shaken. It did not stop him from continuing to speak publicly about his experience. On one occasion, Chris was invited to speak at an event titled "Faithful Witness in a Time of Endless War." It took place on the campus of a Mennonite high school in Lansdale, Pennsylvania, in a small auditorium whose stage was festooned with a drone memorial quilt. The quilt had thirty-six panels. On each one was the name of a person who had been killed by a U.S. drone strike. Chris approached the lectern wearing a brown blazer and a subdued expression. He reached forward to adjust the mic and thanked the event's organizers for inviting him to tell his story. Before sharing it, he asked for a moment of silence "for all of the individuals that I killed or helped to kill."

5

The Other 1 Percent

Christopher Aaron joined the drone program for idealistic reasons. Heather Linebaugh joined it for more practical ones. For her, the
military was not a cause. It was an escape route, a one-way ticket out of
the place where she feared she would otherwise be stuck for the rest of
her life.

Heather was born in Harrisburg, Pennsylvania, where she lived until
the age of six, when her parents' rocky marriage ended in a bitter divorce.
Afterward, she and her twin sister bounced around a string of makeshift
residences with their mother before eventually settling in Lebanon, a
small town surrounded by cornfields and dairy farms. Like Bethlehem
and Allentown, Lebanon had once been home to a thriving steel plant. By
the time Heather's family arrived, foreign imports had decimated the steel
industry, reducing the plant to an unsightly cluster of fenced-off brown-
fields that flanked the railroad tracks running through town. Other than
steel, the town's most celebrated product was Lebanon bologna, a dark,
wood-smoked version of the popular deli meat. One New Year's Eve, a
twelve-foot-long, two-hundred-pound Lebanon bologna was hoisted
into the air and dropped. Another year, a camera crew came to one of the
smokehouses to film an episode of a TV series, showing viewers how the
bologna was made. The series was called *Dirty Jobs*.

A couple of diners, some strip malls, country roads that cut through the
surrounding fields and disappeared into the rural hinterland: this was all
there seemed to be in Lebanon and the adjoining counties. To Heather, an
imaginative child who wrote poetry and developed a precocious interest in

art, the town seemed dreary and lackluster, a provincial backwater where she felt like an oddball. It didn't help matters that she and her sister were mercilessly teased at the local high school by their peers, farm kids who drove tractors and listened to country music. Heather gravitated to the more abrasive sounds of grunge bands and punk rock, music that appealed to her edgy sensibility and to the sense of alienation she felt.

In a different town, from a different family, Heather might eventually have found relief from her isolation by attending a small liberal arts college. In Heather's household, where money was always short, this wasn't an option. After high school, she briefly enrolled at a community college in Lebanon, taking classes in the art program, which she came to feel was a waste of time (the quality of the classes was scarcely better than at her high school). She also got a job as a waitress at a local bar, which was where she met her boyfriend, a navy veteran who was among the regulars. When he invited her to move in with him, Heather leaped at the chance, figuring her luck was turning. "I was like, 'oh, sweet,'" she recalled. A few months later, her boyfriend came home drunk and told her to move out.

The breakup left Heather heartbroken and, even more so, petrified: that she would be trapped in Lebanon forever, cycling through an endless loop of bad relationships and dead-end jobs. It was this fear that led her to drive to a nearby strip mall one morning and edge her way to the counter of a military recruitment center.

Joining the military had not figured prominently in Heather's vision of the future, but she was desperate for a change of scenery and, she hoped, of fortune. Her initial plan was to join the navy, where her ex-boyfriend had served and which she figured would enable her to travel to far-off, exotic places, but there was no one at the navy desk that day. Instead, she struck up a conversation with a recruiter from the air force, who invited her to become a deployable ground station imagery analyst. Heather had no idea what this was, but it sounded faintly glamorous. After scoring high on the Armed Services Vocational Aptitude Battery, she found herself on an intelligence base in a heat-drenched town in Texas, breaking down the feed from Reaper and Predator drones. In January 2009, she was transferred to Beale Air Force Base, an hour north of Sacramento, in the foothills of the Sierras. During her first week there, Heather shadowed a mission in Afghanistan with two staff sergeants, who homed in on a target making his way down a rutted dirt road. "We're gonna take him out," one of them said. Heather peered up at one of the computer screens on the wall of the low-slung trailer they'd squeezed into. She watched the screen flash and then saw the thermal signature of blood oozing from a corpse. It was creepy, but also exciting. "I thought, yeah—we blew something up!" she said.

It fell to imagery analysts to help determine when such strikes were warranted and to provide reconnaissance for soldiers carrying out counterinsurgency missions on the ground. One night, Heather was chatting over Skype with a marine stationed in the area of Afghanistan that she was monitoring. Heather offered him advice on what to buy his girlfriend for Christmas. He thanked her for helping out his battalion. Then he asked her if she had read any of the intelligence documents describing the history and cultural practices of the region they were patrolling. When Heather said no, he sent her some of them, figuring it would enhance her ability to do her job.

The documents Heather received were fairly anodyne—lists of Afghan surnames, descriptions of different dialects and ethnic tribes. Yet when she started reading them, a wave of unease washed over her. Until that point, her knowledge of Afghanistan had been drawn mainly from the jihadist videos she had watched during basic training, chilling footage that underscored the vicious cruelty of the Islamic terrorists with whom America was at war. "I thought everyone there was a member of al-Qaeda," she said. Working as an imagery analyst reinforced this impression. The targets under drone surveillance never had specific names. Come to think of it, Heather never actually saw their faces. All she could make out on the video feed were hazy images of their bodies. Occasionally, Heather joked with her peers that they weren't real people; they were gingerbread men.

In the months that followed, Heather began to notice things she hadn't before—a man patching together the wall of a house, a family gathered around a cooking fire. She also started to wrestle with internal doubts. A lot of U.S. soldiers had died in the missions she had monitored. So had a lot of people whose connection to terrorism she now began to question. In one strike she witnessed, the target appeared to be a man carrying a mortar. The victim turned out to be a woman cradling a small child in her arms.

For a while, Heather dealt with the uncomfortable feelings these thoughts triggered by going to a bar after her shift let out and getting wasted. Eventually, she decided to consult a military psychologist. She told him she was beginning to question the point of what she was doing—even, she added, the morality of it. She wasn't sure the drones always struck appropriate targets, she confessed. In fact, she knew they often did not.

The psychologist called Heather's first sergeant to suggest that she be transferred to a desk job for a while. The request was denied, on the grounds that she was needed on mission. Heather went back to serving as an imagery analyst, and her mood continued to darken. Sometimes after strikes, she would duck into the bathroom, lock the door, and cry. At night, she started grinding her teeth compulsively. One morning, she felt a piercing pain in her jaw; the grinding had cracked a molar. She went back to the psychologist, who, after assessing her deteriorating mental state, put her on suicide watch.

"THE PRESSURE OF ECONOMIC NECESSITY"

Heather Linebaugh began serving in the military at the end of 2008, shortly after the election of Barack Obama. She began serving, in other words, at a moment when the military campaigns launched after 9/11 started fading from the headlines and Americans stopped thinking about the endless wars that were being fought in their name.

For the most part, the public did put these wars out of mind. But there were a few holdouts and exceptions. Sometimes on her way to Beale, Heather approached the entrance to the base and noticed a small band of protesters huddled near the gate. The protesters were mostly older—baby boomers with gray beards and walking canes, clad in fleece jackets and Birkenstocks—and they were trying to get her attention, chanting slogans, holding aloft banners, and, sometimes, clutching baby dolls splattered with fake blood.

Among the regulars at these protests was a woman named Toby Blomé. She lived in El Cerrito, a small city in the Bay Area, in a one-story house surrounded by hedges and festooned with peace signs. MOTHERS SAY NO WAR, read a poster on one of the windowsills of Blomé's home. NO DRONE KILLINGS, NO ARMS SALES! declared a sign on the lawn. Even the front doorbell—RING ME FOR PEACE—confronted visitors with a message. After

I rang it one afternoon, Blomé, a tall, fair-skinned woman in her late fifties, opened the door and invited me inside.

Originally from Southern California, Blomé grew up during the era of the Vietnam War, a conflict she first learned about by leafing through mass circulation magazines like *Time* and *Life*, which published harrowing photos of Vietnamese civilians whose villages had been strafed and burned by U.S. bombs. The pictures made an indelible impression on her, but Blomé did not become politically active until much later. In 2003, she attended a demonstration in San Francisco against the Iraq War. While there, an activist from Code Pink, a feminist peace group, handed her a flyer that summoned women to spend a month in Washington, D.C., to do "antiwar work." Blomé had never heard of Code Pink, but the message in the flyer called to her. After discussing the idea with her husband, she decided to take a monthlong break from her job (she worked as a physical therapist) and go to Washington.

Ever since, Blomé had been organizing peace demonstrations in the Bay Area and beyond, she told me over tea in her living room, which was cluttered with peace signs and protest props, including an assortment of miniature caskets strewn across the floor, each one labeled with the name of a country (Somalia, Yemen, Iraq) where U.S. drones had struck. Draped over a table in the kitchen was a sheet of plastic next to a jar of paint that she'd been using to touch up another set of props: fifteen life-size cardboard figures, representing the victims of a recent drone strike in Afghanistan, that she was planning to haul up to Beale for a protest and vigil the following day.

The protest would be small, drawing no more than a few dozen people, which Blomé attributed to a combination of apathy and disinformation. Unlike during the Vietnam War, Americans were no longer shown images of homes that were burned and destroyed by U.S. bombs, she said with a sigh. They were instead told about "precision strikes" in distant places that few people could locate. For a pacifist who thought every day about the harm these strikes could cause, this was a dispiriting thought. But Blomé refused to succumb to despair. She subscribed to the idea that peace came incrementally through small actions, she told me, judging each protest a success if it changed even one person's mind. Among the people whose minds she most hoped to change were the desk warriors on the front lines of America's new virtual wars. Blomé viewed the drone operators at Beale both as victims—targets of the "heavy-duty brainwashing" that persuaded young people to join the military—and as perpetrators who were responsible for pressing the buttons that could cost innocent civilians their lives. "They are the ones that are doing the killing," she said, "so they have a responsibility." One goal of the protests was to expose active-duty soldiers to critical information that might prompt them to do some soul-searching and, perhaps, experience a change of heart. This is why Blomé and her peers positioned themselves near the base's entrance and confronted the service members reporting for duty with attention-grabbing banners and props. DO THE DRONES HEAR THE CRIES OF THE CHILDREN DYING ON THE GROUND? read a sign unfurled at one demonstration. A lot of the cars that passed by ignored their presence, Blomé acknowledged. But she had met former drone operators who told her they started thinking in new ways "when they saw peace activists outside." At nearly every protest, someone would roll down their window to take a pamphlet or wave, she said.

Given how disillusioned she became, one might imagine that at some point Heather Linebaugh was among those who waved good-naturedly at the Code Pink protesters. Instead, she glared through her sunglasses at them, bristling at their self-righteousness. Far from appreciating their presence, Heather viewed it as an affront, as if she and her peers needed a bunch of peace activists from Berkeley and San Francisco to have their consciences roused.

"They assumed that we don't give a shit, that we're just a bunch of brainwashed, nonhuman robots," she said. "They would say, 'You know you're killing people from across the world—you don't care about it, you have no conscience.' But they didn't know us. They didn't know what kind of shit we had to see; they didn't know most of us wanted to go home and fucking kill ourselves.

"There's a reason that after work we'd all go and get trashed, then talk about how fucked-up mission was this week," she went on. "I would go home and drive past these people protesting and then go have nightmares."

It wasn't just the protesters' blindness to her distress that upset Heather.

It was also the air of superiority she felt they gave off, an impression inflected by differences in social class. The ranks of Code Pink were dominated by educated women from middle-class backgrounds who could afford to devote their time to protesting America's wars without worrying about how to pay their bills or make ends meet: people like Toby Blomé. The ranks of the drone program were filled with people like Heather for whom this was an unimaginable luxury, high school graduates from depressed rural areas and hard-luck towns like Lebanon, Pennsylvania. As during the Vietnam War, when some soldiers returning home felt stigmatized by college students from more affluent families who had secured draft deferments, Heather bitterly resented the judgment of people who had the privilege not to be in her shoes. "I can guarantee that none of you has ever been put in a fucking situation where you have to kill someone or have people that you care about be killed," she said of the Code Pink demonstrators. The protesters were equally blind to the power dynamics within hierarchical organizations like the military, she felt, shouting antiwar slogans at low-ranking enlistees who had little say over the scope of the drone campaign. "They're personally attacking these people who have no control over what's going on," she fumed. "We have no control on that base over what's going on with the drone program."

In fact, some might argue, Heather and her peers had a lot of control. If enough of them quit or became conscientious objectors, it would almost surely have gotten the military's attention, not least because the high burnout rate in the drone program made staffing missions challenging. Unlike young people who had been drafted to fight in Vietnam, moreover, nobody had forced Heather to enlist in the military. She had chosen to do so, joining the all-volunteer military that emerged after the Selective Service ended the draft in 1973. Convention holds that Richard Nixon embraced this change to drain momentum from the antiwar movement, whose campaign of draft resistance became increasingly popular as opposition to the Vietnam War spread. But as the historian Beth Bailey has shown, although Nixon saw the political benefits of ending the draft, the voices that ultimately persuaded him to alter the system were not left-wing college students protesting the war. They were right-wing economists like Martin Anderson, a disciple of Milton Friedman's who headed a White House commission that called for military service to be reframed as a matter of individual choice (Friedman was also on the commission). The free market would do a better job than the state of furnishing America with a professional fighting force, Anderson and his peers argued, a view that struck a chord among conservatives who saw conscription as a misguided form of government engineering and an infringement on individual liberty.

Decades later, this argument didn't just appeal to conservative economists. It appealed to many liberal college students as well, including the Ivy League undergrads who enrolled in the political philosopher Michael Sandel's immensely popular course at Harvard, "Justice." As Sandel noted in his book of the same title, when he asked students in the course whether they favored a draft or an all-volunteer army, nearly all said an all-volunteer army. It would be better for everyone, most agreed, if citizens served in the military by choice rather than compulsion. In the same lecture, Sandel posed another question: Would it be fair to allow wealthy citizens to pay poorer citizens to fight for them in wars? As the students were informed, this was not a hypothetical idea. During the Civil War, hiring substitutes to serve in the Union army was legal and had enabled tens of thousands of affluent Americans—among them J. P. Morgan and Andrew Carnegie—to avoid military service. Nearly all of Sandel's students said this would be unfair, constituting "a form of class discrimination."

Sandel then posed one more question: "If the Civil War system was unfair because it let the affluent hire other people to fight their wars, doesn't the same objection apply to the volunteer army?" The military recruits who served in the all-volunteer army were paid collectively by taxpayers, after all, receiving an array of material benefits (enlistment bonuses, educational opportunities) in exchange for their willingness to serve. In theory, those who elected to serve did so freely. But what if this choice was made disproportionately by citizens who otherwise lacked access to these benefits? "If some in the society have no other good options, those who choose to enlist may be conscripted, in effect, by economic necessity," noted Sandel. "In that case, the difference between conscription and the volunteer army is not that one is compulsory while the other is free; it's rather that each employs a different form of compulsion—the force of law in the first case and the pressure of economic necessity in the second."

In a society where access to education and jobs was relatively equal, the pressure of economic necessity might not determine who ended up serving in the military. In a highly unequal society, it was another story. "Only if people have a reasonable range of decent job options can it be said that the choice to serve for pay reflects their preferences rather than their limited alternatives," Sandel submitted. He went on to quote Congressman Charles Rangel, a Korean War veteran who published an op-ed during the Iraq War that called for reinstating the draft, on the grounds that the enlistment bonuses and educational opportunities offered by the military appealed disproportionately to poor people and people of color who otherwise lacked access to these things. In New York, Rangel noted, "70% of the volunteers in the city were black or Hispanic, recruited from lower income communities."

How typical was this? The Pentagon did not track the socioeconomic status of individual soldiers who had been killed or wounded in America's wars. After the United States invaded Iraq, the scholars Douglas L. Kriner and Francis X. Shen began tracking that status on their own, collecting data dating back to World War II on the socioeconomic conditions in the counties from which military casualties hailed. During World War II, the median family income in communities with high casualty rates and communities with low casualty rates was roughly the same, the data indicated. By the Vietnam War, a gap had emerged, with casualties concentrated in communities where the income was eighty-two hundred dollars lower on average. During the wars in Iraq and Afghanistan—wars fought after the draft was abolished—the gap expanded further, to eleven thousand dollars (in inflation-adjusted dollars). The disparity was equally pronounced when it came to injuries, with citizens from the poorest communities absorbing "fifty percent more non-fatal casualties than the nation's wealthiest communities." Because wounded vets were more likely to return to poorer communities, their residents were far more likely to see and feel the impact of war than their affluent counterparts.

The ideal of shared sacrifice was deeply embedded in American culture, affirmed as far back as the colonial era by the likes of Thomas Paine, who declared that, when it came to answering the call to serve one's country, "it matters not where you live, or what rank of life you hold." In contemporary America, it actually mattered a great deal, Kriner and

Shen concluded. There were "two Americas of military sacrifice," they argued: working-class communities that shouldered the burdens of war, and wealthy ones that were increasingly shielded from these costs. A key feature of this inequality was that it was invisible, their study found, unnoticed by the large number of Americans (including 57 percent of self-described Republicans) who, in a survey they conducted, expressed the belief that sacrifice was evenly spread across all socioeconomic groups. Most of the respondents who subscribed to this view believed that military service was purely a matter of individual choice, unrelated to socioeconomic conditions.

To be sure, the ideal of shared sacrifice had never been honored as faithfully in practice as in theory. During the colonial era, more than two hundred laws were passed that exempted citizens from serving in the militias fighting the British army, measures that "benefited the economically successful and the socially well-positioned," noted Beth Bailey, leaving the burden disproportionately on "the poor and poorly established." But shared sacrifice was not simply a popular fable. During World War II, scores of famous athletes and celebrities served in the military, as did thousands of Ivy League students, including 453 Harvard students and graduates who died in uniform, a figure only slightly below the number from West Point. Military service during World War II was guided by what the historian Andrew Bacevich has termed "Patterson's Axiom," the principle articulated by Undersecretary of War Robert Patterson that in a democracy "all citizens have equal rights and equal obligations" and that the duty of defending the country "should be shared by all, not foisted on a small percentage." After 9/11, this axiom gave way to a new arrangement that mirrored the inequality in the rest of society, argued Bacevich in his book *Breach of Trust*, which pointed out that just one in one hundred citizens has actually served in America's more recent wars. This was the "other 1 percent," Bacevich submitted, veterans whose fortunes seemed inversely related to those of the plutocrats on Wall Street with whom the term "1 percent" came to be associated after the 2008 financial crash. As Bacevich noted, the 1 percent who bore the burdens of national defense rarely overlapped with the 1 percent at the top of the income scale. "Few of the very rich send their sons or daughters to fight," he observed. "Few of those leaving the military's ranks find their way into the ranks of the plutocracy." A retired U.S. Army colonel whose son died in Iraq, Bacevich was unmoved by the expressions of gratitude these elites sometimes showered on veterans at sporting events, thanking them for their service—a service that, according to a poll he cited, ranked third among the "ten worst jobs" in America.

The exemption from shared sacrifice wasn't merely unfair. It was also a key reason "good people" were so disengaged from the wars that other, less affluent citizens fought on their behalf. Like Sandel, the historian David M. Kennedy wondered how different this was from the system of paying substitutes that prevailed during the Civil War. "A hugely preponderant majority of Americans with no risk whatsoever of exposure to military service have, in effect, hired some of the least advantaged of their fellow countrymen to do some of their most dangerous business," he observed, "while the majority goes on with their own affairs unbloodied and undistracted."

Soldiers who risked their lives on the battlefield were at least ac-corded some respect for carrying out this dangerous business. Their wounds and sacrifices were honored and recognized. The recognition rarely extended to cyber warriors who sat at computer terminals and whose psychological and emotional wounds were more hidden. Not long after Heather Linebaugh got to Beale, she was assigned to provide over-watch for a mission in a Taliban stronghold in Afghanistan, alerting the marines on the ground to threats—improvised explosive devices, insurgents plotting ambush attacks—that could be spotted on the cameras affixed to the Global Hawk. The work was stressful, not least because a carefully camouflaged fighter or IED could easily escape the camera's eye. During one shift, a group of marines disembarked from a helicopter and stormed a compound that appeared to be clear of danger. After they entered it, insurgents ambushed them. Heather watched the attack unfold in real time; then she saw one of the marines bleed out.

A few months later, another convoy fell under attack after an IED exploded, igniting a fuel truck that caused more "friendlies" to die. Once again, Heather watched the live feed in real time. At home afterward, she surfed the internet and clicked on a news story about the incident. The article listed the names of some of the soldiers who had been killed, including one who had a wife and young son. When Heather read this, she began to sob.

A week later, at a party for her unit, Heather broke down again, this time in front of her supervisor, who tried to comfort her by reminding her that she was "fighting the good fight." The slogan had been drummed home to Heather during basic training, when new recruits were told their mission was to save lives and to protect America from "terrorists" and "towel-heads." For all her alternative inclinations, Heather had internalized this message. She believed that she and her peers were fighting the good fight. But at the party, the words of her supervisor rang hollow to Heather, who wondered whether the mission—which had ended disappointingly, with no progress made in rooting out the Taliban—had been worth it. "Nothing was accomplished by that convoy," she said later. "Those guys died for absolutely nothing."

Whether innocent Afghans might also have died did not yet cross Heather's mind. "I only felt bad about the guys that maybe we would have saved if we somehow had better technology," she said. "It wasn't out of any sympathy for the so-called enemy. It was out of self-preservation for our people." This began to change after the marine whom she befriended over Skype forwarded her documents about the area in Afghanistan she was surveilling. Like Christopher Aaron, Heather was responsible for conducting surveillance operations rather than coordinating strikes. But what she reported could determine whether a missile would be fired, and it now dawned on her that innocent civilians could die as a result. Sometimes, what disturbed her most was not the strike itself but the aftermath, as the smoke cleared and she watched survivors gather up remains or pick through the rubble. "We can't just bomb someone and fly away," she said. "We have to follow through. The bomb hit and [we'd] wait for it to cool down a little bit, and then you can see the body parts. You can identify, like—that could be the lower half of his body, that could be a leg, and then sometimes you'll stick around and watch family come and get them . . . pick up the parts and put their family member in a blanket."

In addition to grinding her teeth at night, Heather coped with the discomfort these images stirred by talking to her mother, who started calling regularly to check in on her. The phone calls drew them closer, not least because they were both struggling. During one conversation, Heather learned that her mother had been diagnosed with metastatic breast cancer. By this point, Heather had taken to drinking herself to sleep to numb the distress she was feeling. Shortly after learning of her mother's condition, she requested an early hardship discharge so that she could go back to Pennsylvania, both to care for her mother and to alleviate her own agony.

Perhaps sensing she no longer wanted to be there, the military obliged. In March 2012, three years after she arrived at Beale, Heather packed up her belongings and returned to Lebanon, emotionally fragile and, to her mother and twin sister, physically unrecognizable. She had left Lebanon a baby-faced nineteen-year-old, with round cheeks and luminous blue eyes. Now her face was gaunt and angular, the jawline taut, the cheeks sunken, the eyes drawn and ringed by dark circles that made her look permanently sleep deprived. The lack of sleep was caused by the night-mares she'd started having. In the dreams, evil creatures were stalking her. Like the targets in the drone videos, the creatures looked vaguely like gingerbread men, with thick stubby limbs and chubby bellies, and they possessed mysterious powers, including the ability to follow her around without physically materializing. The invisible creatures were jinn, she was convinced, demons made of smokeless fire that appeared in some Middle Eastern folklore and that Heather now imagined coming after her to exact revenge, both by hovering over her and by revealing the source of her anguish to the people she most loved. In one version of the dream, Heather found an anonymous letter pinned to the windshield of her car. The note was addressed to her father, a computer programmer who lived in the town of Carlisle, Pennsylvania, and whom she regarded as "crazy wise." Heather's fondest memories of childhood were of the times they'd spent together, hiking through the snow-draped woods or fixing up the siding on his house. Kindly and nurturing, he was "the person who taught us about moral obligation, how to make moral decisions, how to respect other human beings," she said. "Does your father know what you did?" read the note pinned to her windshield in the dream. "Who the fuck wrote this?" Heather wondered. Then she screamed, realizing that one of the gingerbread men was lurking nearby.

"GRAY, AMBIGUOUS PERSONS"

Working in the drone program was not the only job that could exact such a toll. In 2008, the same year that Heather enlisted in the military, Francisco Cantú began training to work in a different kind of war zone. Its front lines were not in Iraq or Afghanistan but on the southern border of the United States, where agents of the U.S. Border Patrol policed the flow of drugs and migrants streaming in from Mexico and Central America. Before Cantú took the job, his mother, a former park ranger who was Mexican American, tried to dissuade him, reminding him that his own grandfather had crossed the border a century earlier to flee the tumult of the Mexican Revolution. Cantú was undeterred, drawn to the austere beauty of the landscape and to a job he was convinced would enable him to understand the border in a way nothing else could. If he did end up apprehending migrants, he told his mother, his bicultural heritage would be an asset, enabling him to offer them comfort "by speaking with them in their own language, by talking to them with knowledge of their home."

In a memoir titled *The Line Becomes a River*, Cantú described how this belief unraveled during the three and a half years he spent on the border. When he began training for the job, an instructor flashed lurid images of police officers who had been executed by Mexican drug cartels and warned the would-be agents in the room, "This is what's coming." In the field, what Cantú encountered instead were "the little people": mules transporting drugs for narco-traffickers who ruthlessly exploited them; women and children who crossed the border to escape desperate poverty and violence, often at risk to their lives. In the desert, agents sometimes found migrants who drank their own urine to avoid dehydration. They came across others whose rotting corpses were covered in ants. Cantú was good at tracking down migrants and hauling them into custody, but his proficiency came at a cost. He began grinding his teeth compulsively and having strange, eerie dreams, littered with images of dead bodies and of a wolf that hounded him.

As he struggled to make sense of these visions, Cantú consulted various books. Among them was *What Have We Done*, by the war reporter David Wood, about a battalion of Iraq War veterans burdened with moral injury. "Long confused with PTSD, moral injury is a more subtle wound, characterized not by flashbacks or a startle complex but by 'sorrow, re-morse, grief, shame, bitterness, and moral confusion' that manifest not in physical reactions but in emotional responses as subtle as dreams and doubts," the book explained. The concept resonated with Cantú. "One does not have to be in combat to suffer from moral injury," he concluded in his memoir, which was published in early 2018. When it appeared, Cantú braced himself for criticism from his former colleagues in the Border Patrol. Instead, the criticism came from the opposite end of the political spectrum. At readings in cities like Austin, immigrant rights activists showed up to call for boycotting his book. Some labeled him a "Nazi."

Had Cantú written a book glorifying the Border Patrol, the condem-nation might not have been surprising. But the portrait he drew was an unsparing one. In one passage, he described members of the Border Patrol urinating on the personal belongings of migrants. In another, he listed the dehumanizing terms—"scumbags," "POWs" ("plain old wets")—that some agents used to describe the people they rounded up. The activists who appeared at Cantú's readings detected no sign of contrition in these passages, much less any desire to expose the toxic effects of institutional racism. They saw it as evidence of personal depravity and directed their anger at him.

The effort to discredit Cantú unfolded amid the furor sparked by Donald Trump, who assailed undocumented immigrants as "rapists" and "animals" and openly encouraged Border Patrol agents to treat them inhumanely while dispatching Immigration and Customs Enforcement agents to round them up. Cantú's memoir appeared just as the news surfaced that under a policy known as "zero tolerance" more than twenty-seven hundred migrant children, some of them infants and toddlers, had been separated from their parents. Court documents would later reveal that the parents of 545 children, including 60 who were younger than five, had still not been located more than two years later.* In the face of a public outcry, the Trump administration eventually rescinded the family separation policy, but the use of harsh methods of deterrence along Amer-ica's increasingly militarized border continued. In December 2018, two

---

* The age of the victims apparently did not trouble officials like Attorney General Jeff Sessions, who told prosecutors along the border, "We need to take away children."

Guatemalan children died while in the custody of the U.S. Border Patrol. The Trump administration later adopted a new policy, "Remain in Mexico," which forced tens of thousands of Central American asylum seekers to wait in makeshift Mexican encampments as their cases were processed. Against this backdrop, some outraged critics came to see all Border Patrol and ICE agents as irretrievably evil, an "American Gestapo" staffed by jackbooted thugs with blood on their hands.

The outrage was understandable. But before one labels Cantú and other Border Patrol agents "Nazis," it is worth considering the reflections of a writer who experienced the depredations of the actual Nazis. In an essay titled "The Gray Zone," the Italian novelist Primo Levi turned his attention to the division of labor in the death camps, where the most sordid and degrading tasks—sweeping up ashes, participating in lethal selections—were often delegated to prisoners, who performed them in exchange for special privileges (an extra scrap of bread, the hope of being spared death). One reason the Nazis resorted to this strategy was a shortage of manpower. Another was moral. As Levi observed, it wasn't enough for the Nazis to murder their victims. They also wanted to defile them, "to burden them with guilt, cover them with blood, compromise them as much as possible, thus establishing a bond of complicity so that they can no longer turn back."

In Levi's view, this was "National Socialism's most demonic crime," orchestrated to rob the victims of their innocence inside a "gray zone" of coerced collaboration where "the room for choices (especially moral choices) was reduced to zero." How should the "privileged prisoners" who carried out these functions be judged? The more willful collaborators were "the rightful owners of a quota of guilt," Levi submitted, having allowed themselves to become "vectors and instruments of the system's guilt." Ultimately, though, Levi called for judgment to be suspended, advising that we reflect "with pity and rigor" on their desperate circumstances and the unenviable alternatives they faced.

The situation Levi described was extreme and, to some extent, unique. In nontotalitarian environments, the room for moral choices is infinitely greater and the price of refusing to collaborate far less severe. But the "gray, ambiguous persons" who were the subject of his essay, pressed into compromising roles by virtue of their abject circumstances, were not so different from the rest of humanity, Levi took pains to emphasize (their spirit was "mirrored" in all of us, he observed, "we hybrids molded from clay"). "The Gray Zone" was also a meditation on power, which "exists in all the varieties of the human social organization," he noted, and could be wielded in any environment to ensure that defiling tasks were allocated to comparatively powerless groups and individuals. Levi arrived in the camps "hoping at least for the solidarity of one's companions in misfortune"— hoping, in other words, that the experience of powerlessness would fortify the prisoners to band together and resist collaborating with power. He left with the opposite view. The desperation of his fellow prisoners made them more rather than less susceptible to the blandishments of the authorities, Levi concluded, in ways that sometimes blurred the line between victims and perpetrators. As extreme as the circumstances in the camps were, his essay invites us to consider how disparities in power can lead people at the bottom of any social order to serve as "vectors and instruments" of unjust systems: because their lack of power heightens their desire to exercise it; or because they feel constrained by more subtle forces than the ones that the prisoners of the Nazis faced, such as the pressure of economic necessity.

Dirty workers operating under this pressure did sometimes have room to make moral choices. When Harriet Krzykowski learned what happened to Darren Rainey, she could have quit her job. She could also have reported the incident and demanded that the perpetrators be held accountable, as George Mallinckrodt had after hearing that one of his patients had been stomped on. Refusing to collaborate with the authorities might have been risky, but it would not have cost the mental health staffers at Dade their lives, as was the case in the "infernal environment" that Levi described. On the other hand, the fact that some room for choice existed could paradoxically heighten the sense of complicity and self-reproach that dirty workers felt. What prompted Harriet Krzykowski to fall silent was, ultimately, that she didn't want to antagonize security or lose her job. These were good reasons, but were they good enough? She wasn't sure, which was why she kept wondering whether she was a victim of the system or a perpetrator. Her confusion was shared by other staff members I met who refrained from reporting abuse for no better reason than that they wanted to continue receiving their paychecks. In the view of George Mallinckrodt, who lost his job after speaking out about the stomping incident, the duty to protect patients from harm should have superseded this consideration. It was an admirably principled position, one shared by medical ethicists and organizations like Human Rights Watch. Notably, though, Mallinckrodt had less to risk from adhering to such a view. He was a bachelor from a wealthy family, not a parent with young children to support and no safety net, like Harriet Krzykowski. The pressure of economic necessity led Harriet and several of her peers to act more cautiously as they navigated the gray zone at Dade, pulled by conflicting impulses that reflected their precarious circumstances.

Similar tensions prevailed among at least some Border Patrol agents, not all of whom were strangers to the dreams and longings that led migrants to cross the border in search of a better life. In his memoir, Cantú wrote that he mentioned to his mother that nearly half of the recruits at the Border Patrol training academy he attended were, like him, Latino. "Some of them grew up speaking Spanish, some grew up right on the border," he told her. "These people aren't joining the Border Patrol to oppress others. They're joining because it represents an opportunity for service, stability, financial security." In this respect, his classmates were not unique: in 2016, Latinos made up the majority of the U.S. Border Patrol. After Trump was elected and promised to hire more agents, a reporter from the *Los Angeles Times* visited a Border Patrol academy in California's Imperial Valley to talk to trainees. Most of them were Latino. They included people like Michael Araujo, whose own uncles had crossed the border illegally. Why was he joining the Border Patrol? Because of the pressure of economic necessity. "It's a job in a county with the second-highest unemployment rate statewide at 17%," noted the *Los Angeles Times*. "Everyone's kind of interested if you're from around here," Araujo explained. "They know it's one of the few places you can get a good job."

In a study of immigration officers of Mexican ancestry who worked for the Immigration and Naturalization Service, the anthropologist Josiah Heyman discovered something similar. The officers in question had grown up in an impoverished area of the Southwest with a highly segmented labor market, watching their Mexican American parents work various marginal jobs—in discount stores, at packing sheds. In this context, landing a position at the INS was highly desirable; it was a "good job," notwithstanding the social discomfort it could generate. Being Latino could compound the moral hazards of working for an agency like the INS, provoking charges of betrayal and disloyalty. One person Heyman interviewed mentioned that a close friend had told her, "You're arresting my *compadres*." After Francisco Cantú's book appeared, some critics accused him of being a "traitor." The accusation echoed the charge that some Black prison guards heard from prisoners who ridiculed them for working in a criminal justice system that systematically oppressed people of color.

To be sure, not everyone who took a job policing the border did so because they lacked alternatives. Some joined out of conviction and showed no mercy for migrants. In 2019, *The New York Times* published a story about the decline of morale in the Border Patrol, owing to the backlash triggered by Trump's policies. One agent complained that people called him a "kid killer"; another said that he and his colleagues avoided restaurants where people might spit on their food. This was the price of working in an agency responsible for herding sobbing children into overcrowded camps where detainees were denied access to food, water, and medical care. But while a few agents betrayed misgivings about the role they'd been forced to play, far more seemed unconflicted, expressing support for the Trump administration's draconian policies. The *Times* described the Border Patrol as a "willing enforcer" of the crackdown on migrants, an attitude apparent on a private Border Patrol Facebook page where agents used callous, racist language to mock both migrants and their congressional sympathizers. "Oh well," read a post about a sixteen-year-old who died in Border Patrol custody.

The agents who posted such messages deserved no pity. Even Border Patrol agents who did feel conflicted were "the rightful owners of a quota of guilt," not least because, unlike the subjects of Levi's essay, they stood no risk of getting killed for refusing to carry out orders. Guilt was especially appropriate for agents like Cantú, one might argue, a college graduate who could have pursued plenty of other careers. As he acknowledged in his memoir, he *chose* to join the Border Patrol, just as he chose, at least initially, to go along with his share of gratuitous cruelty. Among the people who held this view was Cantú's mother. "You weren't just observing a reality, you were participating in it," she tells him at one point in *The Line Becomes a River*. "You can't exist within a system for that long without being implicated, without absorbing its poison." The fact that Cantú felt remorse afterward likely came as little comfort to the victims, just as it failed to appease Primo Levi. "It remains true that the majority of the oppressors, during or (more often) after their deeds, realized that what they were doing or had done was iniquitous, or perhaps experienced doubts or discomfort," Levi wrote of the collaborators in the camps. "This suffering is not enough to enroll them among the victims."

But as Cantú's memoir makes clear, the cruelty of the system was more than just the sum of individual agents' actions. It was a product of policies formulated by politicians and embraced by many citizens. While activists dismissed ICE and U.S. Customs and Border Protection as "rogue agencies," the agenda they enforced was popular with Trump's base. These agencies did not lack for admirers at the MAGA rallies that helped get Trump elected in 2016 and continued throughout his presidency. Although liberals found the nativism on display at these rallies appalling, many were quick to forget that Trump was hardly the first president to oversee a ruthless crackdown on migrants. Amid the uproar set off by Trump's "zero tolerance" policy, Cantú published an op-ed in *The New York Times* in which he noted that tearing parents away from their children was only the latest, most visible chapter in a much longer story of cruelty and callousness. Its roots could be traced back to the 1990s, when the Clinton administration dispatched Border Patrol agents to cities like El Paso to apprehend illegal migrants. Instead of deterring border crossings, the crackdown made them more dangerous, leading desperate migrants to embark on treks through the desert, which many did not survive. According to the reporter Manny Fernandez, between 2000 and 2016 the Border Patrol recorded more than six thousand migrant deaths.

All of this happened before Trump entered office, as did holding migrant children in unlicensed detention centers, a policy that began under Barack Obama, who deported more people through immigration orders than George W. Bush—indeed, more than all his predecessors combined. These inhumane policies provoked little outrage from a public that, perhaps not coincidentally, was conveniently shielded from seeing or hearing about the less palatable consequences.

"For most Americans, what happens on the border remains out of sight and out of mind," Cantú observed in his op-ed, with "the hostile desert" left to "do the dirty work of deterring crossers, away from the public eye."

No one called Heather Linebaugh a Nazi for serving in the drone program. But plenty of people subjected her to belittlement, she told me when we met one day, a few years after she'd been discharged from the military. Heather was tall and thin, with thick black hair that spilled over her shoulders and an intense, slightly mournful gaze. She was wearing cutoff jeans and an olive-green tank top that revealed a maze of tattoos on her arms, including one she'd gotten on the day she'd processed out from Beale. *À rebours*—French for "against the grain"—it read, a slogan that evoked a side of her personality she'd had to suppress in the military, the rebellious streak that had drawn her to grunge bands as a teenager and that later, at Beale, led her to identify with whistleblowers like Chelsea Manning, the former intelligence officer who leaked hundreds of thousands of classified documents to WikiLeaks after growing disillusioned with the Iraq War. In the eyes of Heather's military superiors, Manning was a traitor. In her own eyes, she was a courageous figure, not least for taking a risk that many soldiers who privately harbored doubts about the morality of what they were doing avoided. This was true of Heather herself, who told me that she briefly looked into filing as a conscientious objector while at Beale. The prospect of ending up like Manning, who was court-martialed and accused of "aiding the enemy," a crime punishable by death, gave her pause. When Manning's hearings began, Heather decided to break her silence by going to Fort Meade to express solidarity for her. The venue she chose was a demonstration outside the courthouse in Fort Meade. On hand were activists from various peace groups, among them Code Pink. Heather approached the area where the speakers were lining up. "Hi, I'm Heather Linebaugh—I was in the drone program," she announced when her turn came, only to be interrupted by a chorus of boos.

Like the activists who showed up at Francisco Cantú's readings to call for boycotting his book, the protesters were less interested in what Heather had to say than in condemning her. The reaction infuriated her. "I wanted to say, 'Fuck all of you!'" she said. The guilt she felt, the nightmares she was having: none of this won her any sympathy. Afterward, Heather wondered what the point of airing her beliefs was. Her doubts resurfaced a few months later, after she decided to air them publicly again, this time in an editorial that she submitted to *The Guardian*. "Whenever I read comments by politicians defending . . . drones," began the op-ed, which Heather composed after a string of fitful nights interrupted by disturbing dreams, "I wish I could ask them a few questions. I'd start with: How many women and children have you seen incinerated by a Hellfire missile?" It continued:

> *What the public needs to understand is that the video provided by a drone is not usually clear enough to detect someone carrying a weapon, even on a crystal-clear day with limited cloud and perfect light. We always wonder if we killed the right people, if we endangered the wrong people, if we destroyed an innocent civilian's life all because of a bad image or angle.*

"It's also important for the public to grasp that there are human beings operating and analyzing intelligence from these unmanned aerial vehicles," Heather went on. "I may not have been on the ground in Afghanistan, but I watched parts of the conflict in great detail on a screen for days on end. I know the feeling you experience when you see someone die. Horrifying barely covers it."

Sharing her perspective could help educate the public about the hidden costs of the drone wars, Heather figured. It would also enable her to vent her grief. About a year after leaving the military, she learned that a member of her former unit had died by suicide in his off-base apartment. The soldier was a former "Airman of the Year" who had been hailed as a model service member. A short while later, another drone analyst Heather knew took his own life. Heather mentioned the suicides in her editorial, along with the "depression, sleep disorders and anxiety" that afflicted her and many of her peers.

The venue in which Heather chose to express these thoughts had not been selected randomly. *The Guardian* was the newspaper through which another whistleblower, Edward Snowden, had revealed that the National Security Agency was secretly monitoring the private email and telephone communications of U.S. citizens. Snowden's revelations sparked a nationwide debate about mass surveillance that spread from the halls of Congress to the boardrooms of telecommunications companies like AT&T and Google. Heather was not so grandiose as to imagine that her article would have a similar effect. Still, she told me, she hoped it could at least spark some reflection and dialogue and help raise awareness about the travails of veterans afflicted with wounds that were less visible but no less debilitating than those of conventional soldiers.

On the day her article appeared online, Heather turned on her laptop, clicked on a link to *The Guardian*'s website, and began scrolling through readers' comments. Within hours, hundreds had appeared. The more of them she read, the more shaken she felt. "Appreciate the honesty, Heather, but you should be in custody awaiting trial," one reader wrote. This was the reaction on the left, among people who needed no convincing that drone strikes were wrong and saw people like her as war criminals. On the right, Heather found herself denounced as a traitor and a crybaby, an "armchair warrior" who had no grounds to complain. Instead of sparking dialogue, her article unleashed blame, a flood of hostile messages that spanned the ideological spectrum. After a while, she shut off her laptop, rattled and distraught. In the days that followed, she avoided going online, after friends warned her about the vicious things being said about her. Eventually, the negative attention died down, giving way to something equally jarring: silence.

The best way to help veterans heal from moral injury was by communalizing it, Jonathan Shay argued in *Achilles in Vietnam*, giving soldiers opportunities to tell their story "to someone who is listening and who can be trusted to retell it truthfully to others in the community." This was the principle behind the ceremony I'd attended at the VA in Philadelphia, where veterans were invited to deliver the truth about war to an audience of civilians who were summoned to listen and to acknowledge their shared responsibility.

But what if society didn't want to listen? What if delivering the truth about war was met with a roll of the eyes or a quick change of subject from people who, when it came to hearing about the wars fought in their name, lacked what Everett Hughes called "the *will* to know"? In the media and the human rights community, the most frequently voiced criticism of the drone program was that it lacked transparency, preventing citizens from knowing the truth. "The public simply did not have the information it needed to evaluate the government's decisions," argued Jameel Jaffer, then the deputy legal director of the ACLU's national-security program. "Overbroad secrecy impoverished public debate and corrupted the democratic process."

Critics like Jaffer weren't wrong that transparency was lacking. As the report published in 2017 by the Columbia Law School Human Rights Clinic and the Sana'a Center for Strategic Studies noted, the U.S. government had been "consistently and excessively secret" about targeted killing operations. In response to these critics, the Obama administration eventually agreed to adopt some limited transparency measures, including a rule requiring intelligence officials to disclose the number of civilians killed by drones in places like Yemen and North Africa. In 2019, the Trump administration eliminated these "superfluous reporting requirements," with barely a murmur of protest from Congress or the public. Trump also eliminated a rule restricting drone strikes to high-level militants. This, too, was met with silence.

The muted reaction underscored how "civilized" drone warfare was—civilized as defined by Norbert Elias, with routine violence tolerated as long as it was discreet and disguised. Compared with ground invasions and conventional bombing campaigns, the kill capacity of drones was trifling, some pointed out. Measured by body counts and collateral damage, this was undeniable. Yet the comparative "humaneness" of drone warfare created a hazard of its own. It enabled U.S. officials "to present warfare as a form of virtue," argued Samuel Moyn, a professor of history and law at Yale, waged by military generals who sought to keep the body count low even as they engaged in surveillance and killing operations "across an astonishing span of the earth" in order to sustain America's global hegemony, a hegemony that few opposed. "The containment and minimization of violence in America's wars, particularly when it comes to civilian deaths, have only made it harder to criticize America's use of force in other countries," Moyn maintained. Precisely because drones and special operations forces carried a lighter footprint and caused fewer civilian casualties than conventional forms of warfare, in other words, questioning whether they should be used at all became, paradoxically, more difficult. It gave America's wars a "moral sheen"—a cleanness—that made the prospect of endless war more palatable.

In fact, the drone program was immune to conventional exposure, the artist and geographer Trevor Paglen suggested. After 9/11, Paglen, a provocateur drawn to clandestine places, began photographing drones and secret U.S. military bases, often by driving down unmarked roads abutting fenced-off areas and attaching a camera to a high-powered telescope. Taken on the sly, Paglen's photos appeared to subvert the logic of official secrecy, documenting the existence of bases like the Desert Rock Airport, a private airfield in Nevada where "torture taxis" ferreted detainees to various stealth locations after 9/11. But the pictures he took didn't actually expose much covert activity. Instead, Paglen set about creating allegories, blurred, ethereal images that underscored the opaqueness of the secret world at which he was pointing his camera. Among the more arresting pictures were photos of the Nevada sky flecked with faint black specks that might have been mistaken for birds. The birdlike objects were Reaper and Predator drones, floating on the edge of a band of milky clouds or veiled by the shimmering glow of a dazzling sunrise. Armed drones had infiltrated the landscape in a way that was all but undetectable, the images implied, subverting our ability to see them clearly or even to notice their presence.

In *Blank Spots on the Map*, a book published in 2009, Paglen argued that when it came to the "black world" of classified defense activity, Justice Louis Brandeis's famous maxim "Sunlight is said to be the best of disinfectants" was little more than a comforting illusion. Far from being amenable to public exposure, secrecy tended to "sculpt the world around it," he posited, creating secret budgets, secret evidentiary rules, secret oaths that employees in the "black world" had to sign that prevented damning information from coming to light. At one point in the book, Paglen told the story of some workers at a secret test site who contracted a mysterious illness that caused their skin to erupt in strange welts and that they suspected resulted from exposure to toxic chemicals. After some of the workers died, a wrongful death lawsuit was filed—a suit in which none of the plaintiffs could be named, because all had signed secrecy oaths (the lawsuit was dismissed, after the government successfully invoked the "state secrets" privilege to prevent any evidence from being disclosed). The "black world" wasn't just a set of obscure bases, Paglen argued, but a vast, ever-expanding storehouse of knowledge and information that no amount of sunshine could penetrate.

It was a dark vision. But as even Paglen acknowledged, the state's ability to keep "blank spots" hidden was not absolute. One of the themes of both his art and his book was that secrecy contained internal contradictions. Among these contradictions was the fact that physical places could not be hidden without announcing their existence in some way. "Blank spots on maps outline the things they seek to conceal," Paglen observed, which is indeed how the secret outposts of the "black world" had come to his attention. Before becoming an artist, Paglen obtained a PhD in geography at UC Berkeley. One day while doing archival research, he noticed that large swaths of land had been redacted from the archive of the U.S. Geological Survey, "blank spots" that piqued his interest. Like the blacked-out portions of government documents that sometimes caught the eye of investigative journalists and historians, hidden objects could inadvertently draw attention to themselves in this way. "More often than not, their outlines are in plain view," Paglen averred.

For all the efforts to keep it concealed, the truth is that the drone program was not so secret. Its existence was hiding in plain sight, in places far easier for ordinary citizens to reach than the covert bases Paglen photographed. To find out how many people had been killed by U.S. drone strikes, one merely had to visit the website of the Bureau of Investigative Journalism, one of several organizations that posted detailed estimates online. As risky as it was for drone operators to speak about their experiences, some nevertheless did. And as difficult as classification restrictions made it to report on the drone program, plenty of enterprising journalists found ways: Jonathan Landay of McClatchy, for example, who drew on top secret intelligence reports to show that missiles fired from drones often struck low-level militants rather than terrorists who posed an imminent threat to national security, as the government claimed. ("I'm thankful that my doctors don't use their definition of imminence," a former air force lawyer who had served as the chief prosecutor at Guantánamo told Landay. "A head cold could be enough to pull the plug on you.") Drawing on another cache of classified documents, Ryan Devereaux, a reporter for *The Intercept*, found that unnamed bystanders made up the vast majority of the victims killed by drones in a campaign in northeastern Afghanistan.

It was true that Americans rarely heard from civilians who lived in places where U.S. drones patrolled the skies, but information about this, too, was not hard to track down. Among the places where it could be found was *Living Under Drones*, a report published in 2012 by the Global Justice Clinic at New York University's School of Law and the Stanford International Human Rights and Conflict Resolution Clinic. The report drew on interviews with more than 130 witnesses and survivors of drone strikes in an area of Pakistan, including Khalil Khan, who rushed to the scene of a missile fired from a drone that slammed into a meeting of suspected militants in the town of Datta Khel. Khan spent the rest of the day gathering "pieces of flesh and put[ting] them in a coffin," he told the researchers, one of dozens of coffins paraded through the streets by grieving villagers after the attack on the meeting, which turned out to be a gathering of tribal elders who had come together to resolve a mining dispute. "They are always surveying us," said one villager quoted in the report. "We are scared to do anything," complained another. The ceaseless buzzing of the eyeless planes circling overhead precipitated "emotional breakdowns, running indoors or hiding when drones appear above, fainting, nightmares and other intrusive thoughts, hyper startled reactions to loud noises." Some villagers were so frightened that they refused to send their children to school. Others avoided crowded places.

Finding such reports was not actually difficult, provided one bothered to look. Many people didn't bother. "The blank spots on the map that Paglen describes have their corollary in the blank spots in the mind and in public dialogue," observed the writer Rebecca Solnit in an essay included in *Invisible*, a collection of Paglen's photographs of the Pentagon's "black world." To judge by the reaction to Heather Linebaugh's article in *The Guardian*, these spots remained blank less because of secrecy than because many "good people" chose not to fill in the details—because they preferred to be kept in the dark.

By the time I met Heather, she had stopped trying to share the truth with a society she sensed didn't really want to hear it. She was focused instead on healing herself—through yoga, through meditation, through an experimental form of massage therapy designed to help people overcome trauma and emotional pain. In this, she appeared to be taking after her father, who, she said, had always sought fulfillment by looking inward rather than seeking validation from others. It was a lesson she'd sometimes struggled to apply to herself, including on the day she had chosen to enlist in the military. That decision was driven not only by her wish to get out of her hometown but also by a desire for external validation. Joining the armed forces would enable her to prove all the people who had mocked and doubted her wrong, she believed, to show that she could make something of her life in a patriotic town where American flags hung from the balconies of many homes and military service was viewed with respect. Among the people whose respect she wanted to earn was her father, whom she went to see before leaving for basic training. When she broke the news to him about her decision to serve her country, she figured he'd be proud. Instead, he offered her a gentle warning. "Just remember that the military's mission is to fight wars and kill people," he said.

PART III

ON THE KILL FLOORS

6

Shadow People

In north-central Mexico, in the state of San Luis Potosí, Flor Martinez grew up with her grandparents in a small adobe house with no electricity and no running water. The house was in the hills, and the hills were beautiful, but Flor's grandparents were poor, and her grandfather was a violent alcoholic. Whenever he got drunk, he would fly into a rage and threaten to kill Flor's grandmother. As a little girl, Flor remembered racing through the house to hide the knives and guns from him. One time, when she was twelve, she crouched behind a chair and watched her uncle pin her grandfather onto a bed to prevent him from stabbing her grandmother.

That same night, Flor learned that her grandmother was leaving San Luis Potosí and that, in two weeks, she herself would be picked up and taken to another town by her mother. The news came as a shock to Flor, who, until this moment, had assumed that her grandmother *was* her mother. "No, no," her grandmother told her, explaining how, shortly after Flor was born, her mother had gone to work as a live-in housekeeper for a wealthy family in another town, leaving her with no time to look after a child of her own. As Flor subsequently discovered, her mother had since started a new family with a man who would soon be taking them to yet another town—a place called Lampasas, in central Texas—to pursue a better life. In the months that followed, after her mother and stepfather departed for the United States, Flor's own life was subsumed by the needs of her two- and four-year-old half brothers, who were left in her care until money could be saved to pay a smuggler to bring them along. To scrounge up food, she would wake at four in the morning, sneak onto a barge, and collect discarded provisions (rotted bananas, moldy tortillas) from a local garbage dump.

It was a perilous existence, but Flor did not feel sorry for herself, convinced her fortunes would eventually turn. Her optimism was tested in the next phase of her upbringing, which began just before she turned fifteen, when a coyote arrived to take her and her siblings to Texas. When they reached the border, Flor clung to a raft that the coyote maneuvered across the Rio Grande (she couldn't swim). Then she heard the coyote holler, "¡La Migra! ¡La Migra!" as helicopters circled overhead, prompting them to turn around. After ducking behind some bushes, they crossed the river again, this time successfully, but the person who was supposed to pick them up and take them to Lampasas never appeared. Flor and her siblings were brought back to Mexico and discarded at a seedy boarding-house teeming with drug dealers and prostitutes. One night, she ventured out to the bathroom and saw an addict shooting up. They stayed there for six weeks, holed up in a squalid room where they slept and ate all their meals until Flor's mother sent more money to the smugglers, who, this time, brought them to Lampasas.

At long last, Flor was reunited with her mother, who lived in a guest-house on the outskirts of a sprawling ranch where her stepfather had gotten a job. The American couple who owned the ranch were welcoming, calling Flor "Sissy" and encouraging her to go to school and learn English. Flor's stepfather was less gracious, informing her that if she lived with them, she would have to pay a price. The price he had in mind was sex. When Flor refused his advances, he was undeterred. "What do you think, your mother is going to defend you?" he would taunt her, reminding her that she'd been abandoned as a child. But although he was a large man who could easily have overpowered her, Flor was not afraid. The experiences she'd survived, and the abuses she'd watched her grandmother endure, had imbued her with a streak of fearless defiance. If her stepfather ever touched her, she would resist, she vowed to herself. Sure enough, one night after the lights went out, she felt her stepfather's hand on her leg. Flor screamed, and her stepfather covered her mouth, but she squirmed out of his grasp and started kicking furiously. The commotion woke her mother, who entered the room and asked what was going on. "Mom, he was touching me!" Flor shrieked. Her mother shot her an icy glare, as though *she* were at fault. Then she slapped her so hard that her head smacked the wall and blood ran from her mouth.

After this, Flor knew that her stepfather was right—nobody would defend her—yet she continued to refuse his advances, hiding a knife between her sheets and her pillow in case he snuck into her bed again. Eventually, her stepfather grew so frustrated that he kicked her out of the house. As she ran off, Flor felt a rush of relief. Finally, she was free, she thought as she fled the ranch on foot. But a few hours later, her legs began to tire. As dusk drew near, she curled up to rest on an embankment near an overpass. Where would she sleep? she wondered as she watched the sky darken and the cars zoom by. Before the last of the light faded, a van slowed down in front of her. The driver rolled down the window and addressed her in Spanish, telling her that he and his Mexican family lived nearby and would take her in.

For nearly a year, Flor lived with the family. During this time, she contacted her mother to tell her where she was. She also asked her to find the address of someone else she'd decided to contact: her biological father. He, too, lived in Texas, she'd heard. She figured she had nothing to lose by reaching out to him. After her mother found an address that matched her father's name, Flor deposited a letter in the mail. A week or so later, a towering man with broad shoulders and a shaggy beard pulled up in front of the Mexican family's house. When Flor reached up to hug him, she could scarcely believe it was her father, because he was so tall. After they embraced, he drove her to his home in Brazos County, a rural area that was home to Texas A&M University, to vast expanses of scrub and farmland, and to a smattering of commercial and industrial enterprises, among them a poultry plant owned by a company called Sanderson Farms.

In his 2018 book, *God Save Texas*, **the journalist Lawrence Wright, a** lifelong Texan, coined a phrase to describe the 1.6 million undocumented immigrants in the Lone Star State. He called them "shadow people" and suggested their existence was as ubiquitous as cowboys and cattle ranches. "One can't live in Texas without being aware of those shadow people," Wright observed. "They tread a line that the rest of us scarcely acknowledge.

At any moment, everything can be taken away, and they are thrown back into the poverty, violence, and desperation that drove them to leave their native lands and take a chance living an underground life . . . The shadow people provide the cheap labor that border states, especially, depend upon. They are not slaves, but neither are they free."

In Brazos County, the shadow people worked various menial, low-wage jobs—as farmhands and dishwashers, as landscape and construction workers. Some also worked at the poultry slaughterhouse owned by Sanderson Farms, which was located in the town of Bryan. A few years after her father brought her to Brazos County, Flor applied for a job there, not under her real name, but as "Maria Garcia," the name on the fake green card that she'd acquired (Flor Martinez is itself a pseudonym). The document raised no eyebrows at the plant, which instantly hired her. According to Flor, a lot of the workers were undocumented at the time, a fact that was no secret to the plant's supervisors, who periodically warned the workers that if they complained, the supervisors would call immigration.

Flor's initial stint at the plant did not last long, not because a supervisor made good on this threat, but because, a few months after she started working there, she overheard that management was looking to hire more supervisors. She passed this information along to a person she thought might be interested—her husband, Manuel, whom she'd recently married after dating for a few years. Manuel got the supervisor job, which created some awkwardness for Flor, both because he knew her real name and because, in the meantime, the two of them learned that his application to become a U.S. citizen had been approved. The news meant that Flor might soon obtain a real green card. When it arrived in the mail, continuing to work at the plant as "Maria Garcia" came to seem foolish, particularly if immigration officials ever *were* called in to check people's papers. After discussing the matter with Manuel, she decided to quit.

Perhaps, Flor thought at the time, she would go back one day and work at the plant under less shadowy circumstances, although another part of her simply felt relief. The job she'd held was in the evisceration department, slicing the glands off the carcasses of chickens suspended on a conveyor belt—the so-called disassembly line—as it rotated by. The cavalcade of decapitated birds was unsightly. Even worse was the smell, a foul blend of chicken excrement and raw viscera that filled the air. She wouldn't miss inhaling that stench for hours on end.

After she quit, Flor spent several years focusing on raising her and Manuel's three young children. When their youngest daughter started school, she began looking for work again. The first position she found was a part-time job mixing salads at a Texas A&M cafeteria. One day, a customer spotted her and, impressed by her diligence and her cheerful manner, asked whether she might want to come work for him. He owned a local Chick-fil-A, where Flor soon began taking orders in English, a language she'd barely studied and had to learn on the fly. Undaunted, she was soon promoted to team leader, then shift leader, then branch manager. The only problem was that, after all this, she was still making only slightly above the minimum wage. Flor thought she deserved better, so she went to talk to the owner who'd hired her. "Oh, sweetheart, you've come a long way," he said, before explaining that she would have to become fluent in English to get a raise.

Line workers at the poultry plant earned between eleven and thirteen dollars an hour—a pittance compared with some factory jobs, but better than anything else Flor could find, which was why, eventually, she put aside her memories of the less appealing aspects of the job and applied to work there again. This time she was assigned to "live hang," where workers hoisted live chickens out of crates and hooked them by their feet onto metal shackles that were fastened to the conveyor belt that circulated through the plant. Once attached to the belt, the birds passed through an electric current (which stunned them), an automated throat slitter (which sliced their necks), and a tank of scalding water (which loosened their feathers). If a chicken somehow emerged from the tank alive, which happened on occasion, a worker wielding a knife would slash its neck manually. The first time Flor saw this, she cried and vowed never to eat chicken again. Most of the time, though, she was in too much agony to think about the chickens. Live hangers had to put sixty-five birds on the belt per minute, a frenetic pace that required lifting the chickens up two at a time, one in each hand, and then immediately reaching down to grab the next pair. For the larger men on the shift, repeating this movement for hours on end was grueling. For Flor, a petite woman with small hands, it was excruciating. After a few days, she could no longer feel her forearms, which were numb with pain. At night, she devoured painkillers to soothe her throbbing neck and shoulders.

Eventually, the pain led Flor to ask her supervisor for a different job. She was switched to "twin pack," joining a crew of workers who slid broilers into plastic bags on the other end of the line. The job demanded far less heavy lifting, but the repetitive strain was worse, not least because the bags were often stuck together, forcing workers to pry them open with their fingertips before maneuvering a chicken inside. Flor's wrists and fingers started to ache, particularly on her left hand. After a while, she went to see a nurse at the plant health clinic. "I can't do this—please help me," she begged. She was transferred again, this time to the debone department, which reduced the strain on her left hand but soon caused her other hand to ache.

The pain was constant, yet the physical distress was not what bothered Flor the most. What upset her even more was the verbal and emotional abuse that accompanied it. The supervisors at the plant never asked her how she was feeling. Instead, they berated her. "You just don't want to work!" one of them snapped. Their only concern seemed to be running the lines at maximum speed, prompting them to bark at the workers as though they were wayward children. Flor was a survivor, and an optimist, but the humiliating tone touched a nerve, reminding her of the way her grandfather used to shout at her grandmother. The most humiliating ordeal of all was requesting to go to the bathroom, which required stepping away from the lines. Although workers were given a thirty-minute lunch break and one other short respite during their shifts, the bathrooms were crowded during these interludes. If they asked to go at other times, they were frequently chastised. Because they were afraid of the supervisors, Flor learned, some of her female coworkers wore an extra pair of pants beneath their work uniforms and, when desperate, wet themselves on the production line. Having witnessed and survived worse bullying, Flor was not afraid, sauntering off to the bathroom without asking for permission when she really had to go, an act of insubordination that prompted the supervisors to glower at her.

The irony was that Flor was herself married to a supervisor, to whom she began to voice complaints at home. "Why do you pressure the workers so much?" she would ask Manuel, pleading with him to advocate on their behalf. He was unmoved. "You're not the only one—deal with it," he would say. The pressure she felt was nothing, Manuel told her, next to the pressure that he and his fellow supervisors felt from *their* superiors, who badgered them at meetings to push the workers even harder.

TORTURED FLESH

I first met Flor Martinez at Our Lady of Guadalupe Hall, a community center across the street from an adobe church in Bryan, Texas. The center was hosting a workshop to educate poultry workers about their rights. A contingent of Guatemalans had come from North Carolina, a group of Mexicans from Arkansas, the home of Tyson and one of the leading poultry-producing states in the country. During breaks, the workers mingled outside, chatting in Spanish and filling their plates with homemade flautas and tamales prepared by one of the workshop's organizers.

Why were so few native-born Americans hired to work in poultry plants? In 2017, the podcast *This American Life* aired an episode in which some residents of Albertville, Alabama, were invited to weigh in on this question. A small town in the northeast corner of the state, Albertville was home to two chicken plants that, like the rest of the poultry industry, had boomed in the 1990s, the decade when chicken was successfully marketed as a low-cholesterol alternative to beef. In places like Albertville, the boom in chicken consumption meant jobs, but the jobs went mainly to Mexicans and Guatemalans, whose arrival fueled resentment among locals wondering why more Americans weren't hired. "There are people out there who want jobs," complained a woman named Pat who'd begun working as a giblet wrapper at one of the plants in the 1970s. "They just quit hiring Americans."

This view was shared by some influential Alabama politicians, among them the then senator Jeff Sessions, an ardent foe of immigration whose perspective on the issue was shaped by the changing composition of the poultry workforce in his home state. ("I really doubt that he would have made immigration his signature issue if not for his experience with the poultry plants in Alabama," Roy Beck, a close friend of Sessions's and the founder of the anti-immigrant group NumbersUSA, told *This American Life*.) At one point, Sessions attended a town hall in Albertville where residents vented their frustration about the immigrants who had flooded into their community. It was enough to convince him that permissive immigration policies enabled foreigners "to take away the few jobs there are, leaving Americans unemployed."

Whether the jobs in question would actually have been desirable to many Americans was open to question. Yet to some extent, the influx of immigrants was what *made* them undesirable. In Albertville as elsewhere, working in a chicken plant became "immigrant work," the status of which was diminished by the hiring of foreign-born workers who exerted downward pressure on the wages and bargaining power of all employees in the industry. The plant where Pat the giblet wrapper worked was a case in point. It had a union, but because Alabama was a right-to-work state, new hires weren't required to join it, and a lot of immigrants didn't, causing membership to drop to 40 percent (it had once been 94 percent). Pat's salary also dropped; at the time the episode of *This American Life* aired, she made $11.95 an hour, roughly half of what she would have earned if her wages had kept up with inflation. In 2002, workers in poultry plants were paid 24 percent less, on average, than the average manufacturing worker. By 2020, they were paid 40 percent less. In theory, the growing popularity of chicken should have tilted the bargaining power to workers' advantage and forced companies to raise wages. The supply of immigrants willing to work for less—desperate workers whom the company preferred, locals like Pat sensed, because they didn't join the union and rarely complained—spared the industry this burden.

This dynamic was familiar to the economist Philip Martin, who studied how the entry of immigrants into certain niche labor markets (picking vegetables, cleaning hotel rooms) made these jobs harsher and less appealing, validating and confirming their unattractiveness—their dirtiness—to native-born Americans. "Employers feel under no compulsion to upgrade dirty jobs as long as immigrant workers are available," observed Martin, "guaranteeing that dirty jobs get less and less attractive to Americans." Immigrants thus acquired a kind of social dirtiness that was tinged with racism and exacerbated by class anxiety as low-skilled American workers feared that more pliant foreigners were displacing them. "It made us all think that we were gonna be pushed out the door," said Pat.

To be sure, Americans with more liberal views of immigration did not see foreign-born workers this way. They saw them as resourceful strivers who did the hard, thankless jobs that no one else would otherwise have done. Liberals were more likely to shower these immigrants with praise than to view them as dirty interlopers. When the jobs in question involved slaughtering animals for mass consumption, however, it was another matter. The killing of animals raised on industrial-style factory farms was, after all, associated with many things—the mistreatment of livestock, the overuse of hormones and antibiotics, the despoliation of the environment—that liberals found abhorrent. Factory-farmed meat was "tortured flesh," as the writer Jonathan Safran Foer argued in his bestselling book *Eating Animals*: carved off the bones of grossly overfed, genetically engineered chickens, cows, and pigs that were crammed into filthy, disease-infested sheds, deprived of sunlight, and subjected to untold cruelty and suffering, all to maximize the profits of a handful of giant corporations.

To eat this meat was to be complicit in the torture, Foer's book implied, a message that resonated with a growing number of health- and eco-conscious consumers who preferred to buy organic meat that came from family farms or opted to become vegetarians. But if consuming factory-farmed meat was deplorable, what did this say about the people who hacked apart the animals, the live hangers and tendon cutters who worked for the giant corporations and were directly involved in the killing process? To the extent that they appeared in exposés of the meat industry, these workers tended to be depicted less as resourceful strivers worthy of admiration than as callous brutes. In a section of *Eating Animals* titled "Our New Sadism," Foer described a Tyson facility where workers "regularly ripped off the heads of fully conscious birds." He described another facility—a KFC "Supplier of the Year"—where "chickens were kicked, stomped on, slammed into walls, had chewing tobacco spit in their eyes." If you worked in the industry, you were likely to turn into a vicious torturer, such stories suggested, although, to his credit, Foer acknowledged that the workers themselves were often severely mistreated as well. Such acknowledgments rarely appeared on the websites of organizations such as People for the Ethical Treatment of Animals, which posted undercover videos of "sadistic workers" inflicting abuse on farm animals and advocated charging the perpetrators with criminal felonies. "The blasé attitude towards unbearable suffering and the outright sadism that you can see on this video turns up again and again and again when one of these hellholes is exposed," one PETA video and blog post declared.

Both on the right and on the left, then, albeit for very different reasons, the people working in America's slaughterhouses are likely to be viewed disparagingly. To be seen as dirty, which is how people whose jobs bring them into direct contact with the flesh and blood of animals have long been seen in many cultures. "In Tokugawa-era Japan, butchers were categorized among the *eta*, or unclean people, and had to live and work in segregated parts of cities," notes the historian Wilson J. Warren. "In India, people who worked with dead animals were part of the untouchable caste." The condemnation was less sweeping in countries such as France and England, but direct involvement in the killing of animals nevertheless carried a moral taint. In his influential treatise *Some Thoughts Concerning Education*, published in 1693, the philosopher John Locke noted that butchers tended to be excluded from juries, on the grounds that "they who delight in the suffering and destruction of inferior creatures, will not be apt to be very compassionate or benign to those of their own kind."

More than two centuries later, in 1906, a searing exposé of the meat industry appeared in America. By the time Upton Sinclair's *The Jungle* was published, the era of the village butcher, a skilled craftsman who catered to the members of a local community, had given way to the era of large meatpacking companies that capitalized on the invention of refrigerated railroad cars to transport meat from distant farms to centralized slaughterhouses in big cities like Chicago. (They also capitalized on the absence of antitrust laws to stifle competition and maximize their profits.) It was in a Chicago slaughterhouse that Sinclair's novel was set. Although it was a work of fiction, Sinclair spent seven weeks in the stockyards of Chicago before writing the book, which drew its force from its harrowing realism. Like Foer's *Eating Animals*, *The Jungle* described the mass slaughter of livestock in grisly detail, "a very river of death" presided over by workers who emerged from their shifts spattered in blood. Unlike in Foer's case, Sinclair's primary goal was not to draw attention to the mistreatment of animals. It was to dramatize the plight of the workers, which Sinclair, a socialist, hoped to change. "The novel I plan would be intended to set forth the breaking of human hearts by a system which exploits the labor of men and women for profits," he informed his publisher. Among the brokenhearted was *The Jungle*'s protagonist, Jurgis Rudkus, a Lithuanian immigrant who comes to Chicago to pursue the American dream. After his elderly father dies from a lung infection contracted while working at a meatpacking plant, Jurgis cannot afford to give him a proper funeral. He ends up getting fired after sustaining an injury at the plant, where workers are routinely exposed to unsafe conditions and denied basic amenities, such as heat during winter and access to toilets. "There was not even a place where a man could wash his hands," wrote Sinclair in one of many passages that suggested that working in the meatpacking industry was not only dangerous but befouling.

First serialized in a socialist newspaper, *The Jungle* was a popular sensation, shocking readers horrified by its descriptions of dead rats and tubercular beef that was ground up and sold to unsuspecting consumers. Sales of meat declined sharply. President Theodore Roosevelt invited Sinclair to lunch at the White House. Sympathetic to the idea that the meatpacking industry was a nefarious trust—if not to the cause of socialism— Roosevelt soon dispatched investigators to probe conditions in Chicago's stockyards. The investigation helped spur passage of the Meat Inspection Act and the Pure Food and Drug Act. For Sinclair, a twenty-seven-year-old writer whose previous books had been critical and commercial failures, the reaction was intoxicating. But it was also sobering. *The Jungle* made Sinclair the most famous muckraker of his generation, but it did not allay his concern that once the fear of eating contaminated meat abated, the big packing companies would go back to exploiting their workers as ruthlessly as before. What roused the indignation of his readers was not the mistreatment of these workers but the risk of eating rancid meat, Sinclair ruefully concluded. "I aimed at the public's heart, but by accident I hit it in the stomach," he later remarked.

Upton Sinclair died in 1968, by which point concern about the working conditions in the stockyards of Chicago had faded from the headlines.

Yet in the intervening decades, the conditions in the industry improved, not because the public demanded this, but because workers did. A major force behind these improvements was the United Packinghouse Workers of America, which successfully organized an industry in which racial and ethnic divisions were deliberately stoked. At the turn of the century, the workforce in meatpacking plants had been dominated by eastern European immigrants. By 1930, one-third of the workers in Chicago's stockyards were African American. Many Black workers were initially recruited into the industry as strikebreakers and were given the harshest, least desirable positions. During and after shifts, a "social and cultural apartheid" prevailed that prevented workers of different racial backgrounds from mingling, much less from linking arms on picket lines. The UPWA went to great lengths to bridge the divide, holding racially integrated rallies, integrating the bars and taverns near the stockyards, and encouraging Black workers to serve as shop stewards. Its efforts did not go unnoticed by publications such as *The Chicago Defender*, which praised the union's struggle to "defeat prejudice." In the decades after World War II, its efforts also began to pay off, resulting in industry-wide wage scales that were 15 percent higher than the average for all manufacturing jobs.

Working in a slaughterhouse was still not easy. It was a dirty, difficult job, associated with an activity—killing animals—that aroused widespread disgust. But for several decades, meat packers were able to earn a respectable living and to exercise their collective bargaining rights, making the story told in *The Jungle* seem obsolete. The improved conditions lasted until the early 1970s, when a company called Iowa Beef Packers pioneered a new production model. Instead of locating its plants in cities, IBP situated them in rural areas—closer to farms and cattle ranches, which could reduce transportation costs. The areas in question also tended to be hostile to unions, which lowered labor costs. When a strike broke out at IBP's flagship plant in Dakota City, Nebraska, the company responded by importing Mexican strikebreakers. The move was part of a new "low-wage strategy" that the company implemented to give it an advantage over its competitors. Soon enough, IBP's competitors began to emulate the strategy, bringing the good times for meatpacking workers to a halt. By 1990, wages in the industry were 20 percent lower than the average for all manufacturing jobs. Meanwhile, the injury rate soared as a once-stable job was radically de-skilled and transformed into an increasingly dangerous, temporary one.

Lacerations, torn muscles, distended fingers: the workers in America's slaughterhouses had plenty of tortured flesh of their own. "My scars are many: on my hands, arms, heart, mind and soul," one meatpacking worker said of the brutal conditions he endured. "I have learned that I am nothing to any packer but a fucking piece of dirt." In many slaughterhouses, the turnover rate exceeded 100 percent *annually*. How could plants in rural areas find enough healthy bodies to replace the workers they churned through? Without help from lobbying organizations such as the Essential Worker Immigration Coalition, which pushed for allowing more low-skilled immigrants to enter the country, they might not have. Among the lobby's members was the American Meat Institute. Meatpacking companies sometimes enlisted third-party contractors to bring immigrants and refugees from war-torn nations like Sierra Leone into the United States. By the 1990s, an estimated one-fourth of the industry's workforce consisted of undocumented immigrants whose marginal status made them far less likely to stand up for—or even to know—their rights. In 2005, a Human Rights Watch report revealed that basic rights were systematically violated in industrial slaughterhouses and that these violations were inextricably related to the industry's reliance on immigrant labor. "All the abuses described in this report—failure to prevent serious workplace injury and illness, denial of compensation to injured workers, interference with workers' freedom of association—are directly linked to the vulnerable immigration status of most workers in the industry and the willingness of employers to take advantage of that vulnerability," the report found.

As we've seen, dirty workers sometimes hail from poor, isolated backwaters in America—the "rural ghettos" where many of the nation's prisons are located, for example. In the case of poultry slaughterhouses, many are from other countries, "shadow people" hired to work in an industry that plays an increasingly central role in America's food system and diet. Between 1960 and 2019, the per capita consumption of chicken in the United States more than tripled, surpassing both beef and pork to become

America's most popular meat. By the latter year, more than twenty thousand fast-food chicken franchises had opened for business in the United States, part of a thirty-four-billion-dollar industry. The "shadow people" stood at one end—the invisible end—of a chain that led from America's factory farms through its industrial slaughterhouses to the frozen nuggets sold in supermarkets and the order windows at Chick-fil-A and KFC. As Everett Hughes might have noted, these workers solved a problem for society, performing a distasteful job that someone had to do to keep the customers at these franchises satisfied but that few Americans had the stomach or wherewithal for.

"PLANTATION CAPITALISM"

I heard this sentiment again and again at the workshop in Bryan, Texas, where I met Flor Martinez. A Guatemalan worker named Juan told me that at the poultry plant in North Carolina where he had worked for many years, white people would sometimes come in, work until the lunch break, walk out, and never be seen again. "They say, 'I'm not crazy enough to stand here all day,'" he said with a chuckle. Like the other workers on hand, Juan had come to the workshop at the invitation of the Centro de Derechos Laborales (Center for Labor Rights), a pro-worker nonprofit organization based in Bryan that sought to help immigrants defend their rights. About halfway through the proceedings, after a lunch of refried beans and fresh tamales was served, a group of workers came forward to perform a skit. Dressed in hairnets, smocks, and orange latex gloves, the workers stood side by side behind a waist-high table that they pretended was the production line in a poultry plant. As they simulated various tasks, moving their hands quickly while repeating the same motions over and over again, a male supervisor stood watch, periodically shouting, "¡Vamos!" "¡Rápido!" At one point, one of the female workers asked to go to the bathroom. "Necesito ir al baño," she said.

"¡Aguantarse!" the supervisor shouted, meaning "Hold it in."

"No puedo esperar," the woman responded. "I can't wait."

"¡Aguantarse!" he roared.

Afterward, I spoke with one of the workers who had taken part in the performance, a woman I'll call Regina (she did not want her real name used). Originally from Mexico, she was in her mid-fifties, with brightly polished fingernails and an impish smile that vanished when she started talking about her experience at the poultry plant in Bryan, where she worked for two years. By the time she left, she had developed carpal tunnel syndrome in both arms. She'd also sustained a hip injury when a worker carrying heavy boxes accidentally rammed into her. Performing the skit had been painful, she said. But the source of the pain was less physical than emotional, reminding her of the times she'd stood on the line needing to go to the bathroom and been ordered to wait. This happened so frequently that she developed a bladder problem, she said. Eventually, she started wearing a sanitary napkin to work in case she had to wet herself on the line.

"My logic would tell me that I had the right to go to the bathroom, but I did not want to lose my job," she said with downcast eyes. When I asked her how taking such precautions made her feel, she fell silent. "Sad, mad, impotent," she said. Then she cried. A divorced mother with several children to support, she obeyed the supervisors who humiliated her only because she needed money and feared losing her job, she told me. "Necessity makes us work at that place," she said after wiping her eyes.

A few weeks later, I came back to Bryan to interview several other current and former line workers at the poultry plant. All of them were Mexican immigrants. All were women with children and families to support. When I asked why they worked at the plant, all relayed a variation of what Regina had said: because they needed the money and because it paid better than flipping burgers or cleaning toilets at a motel, the only other work they could find in Brazos County, where good jobs were hard to come by, especially for Latinos, among whom the poverty rate was 37.4 percent. As they recounted their experiences, all but one of the women wept openly.

As the tears suggested, the workers at the plant didn't just feel mistreated. They felt degraded and demeaned, a sensation familiar to the subjects of *Scratching Out a Living*, an ethnographic study of the poultry industry written by the anthropologist Angela Stuesse. Until 1945, poultry production, unlike pork and beef, was a small-scale industry based mainly in southern states like Mississippi, where Stuesse did her fieldwork and where, prior to World War II, the workforce consisted mainly of white women.

By the 1970s, the white women had been replaced by African Americans, whose entry into previously segregated plants prompted many whites to leave. By the early years of the twenty-first century, the industry's Black workers found themselves working alongside Latino immigrants in what had suddenly become a highly profitable business (in 2020, poultry and eggs were Mississippi's top agricultural products, generating $2.8 billion in revenue). Reflecting on how few of the profits trickled down to the workers, Stuesse characterized the poultry industry as a system of "plantation capitalism," featuring a labor force dominated by people of color who toiled under dehumanizing conditions for companies whose owners and senior managers were overwhelmingly white. As on plantations, the workers were subjected to brutal treatment by supervisors who exercised total control over them. As on plantations, the grueling work left both physical and emotional scars. Working in a poultry slaughterhouse "causes more than bodily pain," Stuesse concluded. "Chronic abuse at the hands of superiors also injures the spirit, threatening workers' sense of dignity, self-worth, and justice."

Historians of slavery might take issue with the term "plantation capitalism." Yet as I listened to the workers at the poultry plant in Bryan describe the indignities they endured, the phrase kept coming to mind. "All they lack is a whip for them to hit you," said Regina, snapping her fingers to imitate the supervisors who ordered her around. Others spoke of being treated like "machines" that became disposable the minute their bodies started breaking down. This was the view of Libia Rojo, a friend of Flor Martinez's whom I met on my second trip to Bryan. She'd worked at the plant for eighteen years, an experience that had left her with a damaged shoulder, an injured wrist, and a lame right arm that hung limply at her side. The injuries had recently forced her to stop working, she told me. Now she feared that no one would ever hire her again.

I heard a similar story from Juanita, an undocumented immigrant who lived in a trailer park in Bryan with her husband and their three young children. Inside the trailer, the shades were drawn and the air was stifling, moistened by the steam billowing out of two large pots simmering on the stove of a tiny kitchen. When not sprinkling salt or slicing plantains into the pots, Juanita attended to her children, among them a fretful toddler whose cries for something to eat were temporarily silenced with a bottle of Nesquik. The two other children sat slumped on a threadbare couch in the corner of the room, watching television. The cramped quarters and multiple mouths to feed explained why, when Juanita first started working at the poultry plant, she disregarded the warnings of older workers who told her, "You're young—you should not work here." The job paid $12.20 an hour, which seemed like a fortune to her, and included health benefits (though workers had to pay one-fourth of the cost out of pocket). "At the orientation, I thought, wow," she said.

Not long after she was hired, Juanita slipped in the bathroom at the plant because the floor was slick with grime. She broke her fall with her left hand, which soon started hurting. She went to see the plant doctor, who wrapped her wrist in a bandage and told her she needed to limit her activities. The job she was subsequently assigned—packing chicken breasts onto trays at breakneck speed—was incompatible with this directive, which won her little sympathy from her supervisor, who she began to suspect was plotting to get rid of her. On one occasion, she arrived at her shift two minutes late after going to the nurse to get bandages for her hands, which were particularly sore that day. "You're always late; you do not pack enough!" the supervisor scolded her. "I cannot permit this." When she started crying and asked to go home because the pain was agonizing, he threatened to fire her. "I don't care how you feel: you stay or I fire you," he snapped.

Later, while she was working in another department, the chemicals used to disinfect raw chicken accidentally splashed into Juanita's eyes. The tearing and irritation got so bad that the nurse at the plant sent her to see a specialist in Austin, who told her she needed surgery. For this and subsequent appointments related to her eyes, Juanita told me she had to pay out of pocket. When she came back to work, her eyes continued to bother her, but the nurse at the plant determined that the condition was unrelated to the job. At one point during our conversation, Juanita pulled out a folder filled with medical records. She showed me a physician's report given to her by the company in which the nurse wrote, "No work related injury, is allergy." Juanita glared at the document. "I have never had allergies," she said. She fished out another medical document, from the third party that administered her health plan at the plant. It was an "Adverse Benefit Determination" that concluded the injuries to her left hand, left wrist, and shoulder were also not work related.

"They wanted me to leave the job; that is what they mainly do when you get injured," said Juanita, who told me the only thing her health insurance seemed to cover were cheap ointments and bandages dispensed at the plant clinic. Eventually, Juanita did leave the job. When we met, she was unemployed and unsure if she would ever be able to work again, both because she was undocumented and because of her physical impairments. Although her eyes had stopped tearing, her vision was not as sharp as it used to be, especially when exposed to bright sunlight, which was why the shades in the trailer were drawn. Sanderson Farms, she said, treated her "like a disposable piece of trash."

MISSIONS IN TRANSPARENCY

The day after visiting the trailer park where Juanita lived, I drove to the Sanderson Farms plant in Bryan to see if I could talk to some managers and supervisors about how workers on the disassembly lines were treated. When Upton Sinclair composed *The Jungle*, outsiders curious about such matters could visit a slaughterhouse simply by walking through the streets of cities like Chicago and looking around. "They say every Englishman goes to the Chicago stockyards," one of these outsiders, Rudyard Kipling, wrote in his 1899 *Letters of Travel*, which included a vivid description of pigs and cows slaughtered in livestock pens that stretched on for blocks. "You shall find them about six miles from the city," wrote Kipling, "and once having seen them, you will never forget the sight."

More than a century later, America's slaughterhouses had gone the way of prisons, relocating to the "unobtrusive margins" of society, both to take advantage of business-friendly environments and to be removed from sight. The Sanderson Farms plant was located in an industrial park, off a freeway, at the end of a winding road bounded by a metal gate. PRIVATE PROPERTY PAST THIS POINT, a sign above the gate announced on the rainy, overcast day I visited. After following the road up a gently sloping hill, I reached a parking lot marked off by another security gate. In the near distance, two rain-doused flags—one with fifty stars, the other with just one (the flag of the Lone Star State)—sagged on a pole by the entrance to a massive brick building. As I walked over to the security gate, a semitruck stacked with wooden crates encased in mesh wiring rolled by. The crates, which had arrived that morning full of live, squawking birds, were the only indication that what took place behind the walls of the building was the mass slaughter of animals. They were now empty and spattered in drizzle.

Sanderson Farms, which owned and operated the plant, is one of the world's largest poultry producers and the only Fortune 1000 corporation in Mississippi, where the company was founded in 1947. A family business that started out supplying chicks and feed to neighbors in the town of Laurel, the company went public in 1987. In the 1990s, it expanded its operations into other states (Georgia, Louisiana, Texas), places with weak unions and, no less important, permissive environmental laws. The plant in Bryan opened for business in 1997. Fifteen years later, Environment Texas, a nonprofit group, published a report citing it as the state's leading polluter of water, releasing 1.2 million pounds of toxic discharges into the state's creeks and rivers. This was not unusual. According to a report published in 2018 by the Environmental Integrity Project, the typical slaughterhouse discharged 331 pounds of nitrogen a day, roughly the amount contained in the untreated sewage of a town of fourteen thousand people. Some poultry plants routinely violated local pollution limits, the report found, violations that often went unpunished. Many other plants were located in states with lax regulations "that allow them to discharge far more pollution." If "good people" living in affluent communities weren't bothered by this, it's perhaps because most of them were shielded from the consequences. As the Environmental Integrity Project noted, slaughterhouses were disproportionately located in isolated areas "with a high percentage of Latino and African American residents" and a large number of residents "living beneath the poverty line," communities that "can least afford to lose their drinking water supplies and natural resources."

The dirty by-products (blood, fecal waste) of the meatpacking industry leached into the streams and rivers of the same communities where the people who did the dirty work inside the slaughterhouses lived. Places like Bryan, where, according to the Texas Education Agency, 74 percent of the children in the local school district were economically disadvantaged in 2015.

Like many poultry companies, Sanderson Farms favored a business strategy known as vertical integration, whereby it owned everything from the feed mills and hatcheries to the trucks that delivered the chickens to slaughter. Pioneered by Tyson, this strategy enabled the big companies that dominated the meat and poultry market to reap massive profits. As the reporter Christopher Leonard shows in his book *The Meat Racket*, it was far less beneficial to rural communities and to the contract farmers who raised chickens and hogs. The farmers were paid based on secret formulas, written by the companies, that often left them on the verge of bankruptcy. Those who complained about the terms were often put out of business (the companies, which owned the chicks, would simply stop supplying them). As Leonard notes, rural Americans had coined a term to describe what happened to their counties as companies like Tyson expanded their influence: "They have been *chickenized*."

Also like its competitors, Sanderson Farms took pains to cultivate a positive image. In 2018, as part of this effort, the company launched a "truth-telling" ad campaign to promote its new "Mission in Transparency." "We believe if we are transparent and tell our customers and consumers more about why we do what we do, they will not only have a better understanding of poultry production, but they will also be able to feel better about what they are feeding their families," announced Hilary Burroughs, director of marketing at Sanderson Farms.

The emphasis on transparency had not filtered down to the guards at the security gate of the plant in Bryan, who told me I could not enter the premises after I asked to go inside. One of the officers handed me a slip of paper with the plant's phone number on it, explaining that all visitors needed to get permission before entering. I went back to my car and called the number, reaching an office worker who gave me another number—for Mike Cockrell, the CFO of Sanderson Farms, who was based in Mississippi. I dialed it and spoke to Cockrell's secretary, a friendly woman with a thick southern accent who took down my name and explained that Mr. Cockrell was in a meeting. She said she would try to get his attention and that I should call back.

An hour later, I called back, explaining that I was outside the plant and waiting for permission to enter. Mr. Cockrell was still in a meeting, she told me. After another hour passed, I tried again. Mr. Cockrell was still busy, she said, and would be tied up all afternoon with phone calls and emails, making it impossible for me to get into the plant that day.

In the weeks that followed, I called Sanderson Farms several more times, repeating my request to visit the plant in Bryan. I also asked for an interview with Cockrell. Neither the interview nor my request for a visit was granted. This was hard to square with the company's Mission in Transparency. It was not hard to square with the secrecy for which the meat and poultry industry was known. In a number of states, it was actually a crime to record a video or take a photograph inside a meat or poultry slaughterhouse, thanks to so-called ag-gag laws passed with the help of the industry's lobbyists. (Some of these laws were subsequently struck down by judges who determined that they were unconstitutional.)

Eventually, I did receive a statement from Sanderson Farms about its workplace policies, a two-page press release that had originally been issued in response to a report published by Oxfam America. Titled "No Relief," the Oxfam report focused on the denial of bathroom breaks to workers in poultry slaughterhouses, a problem it portrayed as pervasive. Although Sanderson Farms had declined to address these allegations when initially contacted by Oxfam's researchers, once the report appeared—and received some press coverage—it wasted no time refuting its validity. "The Company does not deny any person the use of restroom breaks," asserted the press release I'd been sent. "Sanderson Farms strictly follows Occupational Safety and Health Administration (OSHA) standards stating that restroom facilities must be available to employees upon need."

"Sanderson Farms' most valuable assets are our employees," the press release went on, "and we treat them with dignity, respect and the utmost appreciation for their dedicated work."

At one point, Flor Martinez was given an opportunity to see what would happen if workers on the disassembly line actually were treated with dignity and respect. It came when she was assigned to serve as a floor worker at the plant in Bryan. The job was a hybrid position, she told me, combining line work with some supervisory duties—duties that Flor tried to dispense with compassion. If a worker on the line looked exhausted, she would replace her so that she could rest. Instead of shouting, she smiled and offered encouragement.

The approach went over well with Flor's coworkers, who responded by breaking the record for the number of birds processed in a seven-hour shift on her third day on the job, she told me with pride. It did not go over as well with her superiors, who berated her for being too lenient. One time, a supervisor told her to give each worker one pair of gloves and, if anyone asked for another pair, to say there were no more. "But we got a full box in the office," Flor objected. "These are people—they need the supplies." The supervisor was adamant. "Flor, please—just do what I say," he demanded.

Watching the supervisors browbeat the workers, Flor would have flashbacks to her childhood, when she'd stood by helplessly as her grandfather intimidated her grandmother. The job was particularly demeaning to women, she felt. It was women who risked getting bladder infections when denied bathroom breaks. It was women who were subjected to sexual harassment. ("Move it, move it like you did last night!" supervisors sometimes shouted at female workers on the lines.) Addressing such harassment was among the goals of the workshop at Our Lady of Guadalupe Hall, which opened with a presentation delivered by two attorneys from Austin. The lawyers explained what qualified as sexual harassment and how employees could contest it, as had been happening with growing frequency in the media and entertainment industries, where women inspired by the #MeToo movement had begun to expose the powerful men who had abused or assaulted them. To sit in on the workshop in Bryan, which took place in the fall of 2018, about a year after the sexual predations of the Hollywood producer Harvey Weinstein were first exposed, touching off a cascade of similar revelations, was to appreciate how quickly the #MeToo movement's impact had spread. But the workshop also underscored the barriers that the movement faced. It was hard enough for Hollywood actresses and female news anchors with access to lawyers to file complaints, given the retaliation and public shaming that could ensue. It was immeasurably harder for Latino immigrants in an industry where workers feared being subjected to retaliation for voicing *any* complaints. And retaliation from employers was not the only problem. After the attorneys from Austin spoke, a discussion unfolded, during which a worker from North Carolina told the story of a woman at a poultry plant who complained to management about a supervisor who had repeatedly harassed her. When the woman's husband found out, he was livid—not at the supervisor, but at *her*. The gender dynamics on the disassembly lines all too often mirrored the gender dynamics in workers' homes and families, the worker said, leaving female workers with few safe places to turn.

When Flor turned to her husband, Manuel, to voice her frustration about the harsh treatment she was expected to dispense as a floor worker, he told her that she needed to stop feeling so much and think more like a supervisor. To think like a supervisor meant to care only about the bonuses you'd get for processing more chickens, Flor concluded. When Manuel came home at the end of the year with his bonus, he would show her the check, expecting her to be happy. Flor did appreciate the money, which enabled them to cover their expenses and, eventually, to move out of the trailer home in Bryan, where they'd lived for many years, into a two-story house in neighboring College Station. But over time, she began to feel bad about the bonus that Manuel brought home—and to see the check as dirty money.

"I'd say to him, why did you move the speed of the lines? Why are you killing the people—you're killing us!" she recalled. "And I wouldn't say Sanderson Farms is killing us. I would say *you* are killing *me*. Why are you doing this?

"He had the power; he had the power to go up there," she said.

By "up there," Flor meant the meetings where supervisors conferred with senior managers. There was no excuse not to speak up for the line workers at these meetings, she felt. Yet as Flor acknowledged, the real power at these meetings did not belong to the supervisors. It belonged to the managers who set the production quotas that determined how fast the lines ran—managers who addressed the supervisors in a tone that was equally belittling, Manuel would tell her. The supervisors were merely doing the dirty work for high-ranking corporate executives, in other words, while people like Joe Sanderson reaped the benefits.*

---

* In 2017 and 2018, Sanderson was the highest-paid CEO in Mississippi, earning more than ten million dollars.

"MATTER OUT OF PLACE"

Flor and the other workers I spoke to were convinced that the workers on the kill floors could be mistreated because management saw them as docile immigrants who would never dare to stand up for themselves. "That's why they do it, because they think we can't defend ourselves because we don't speak English well," said Libia Rojo. It was not easy to defend yourself when you had few other options and felt socially invisible, several workers told me. Yet the truth is that the workers were not invisible. They were highly conspicuous, an imagined "other" inveighed against frequently in Brazos County—a county that, like the rest of Texas, both relied on "shadow people" and vilified them. In 2017, Texas's governor, Greg Abbott, signed SB 4, a law that banned municipalities in Texas from serving as "sanctuary cities" for undocumented immigrants and required law enforcement to honor ICE detention requests. On my way to interview one of the workers at Sanderson Farms, I tuned in to a local radio show where the subject of immigration was discussed. The host of the program was delivering a harangue about how the Latino immigrants inundating Texas "hated America" and threatened its heritage and values. The host did not say the immigrants were dirty, but he might as well have, casting them as intruders who could not be absorbed into the body politic without corroding and polluting it.

All the workers I interviewed told me they had heard more and more such talk during Donald Trump's presidency. It was something new and ominous, they said, fueling hatred and violence.* Yet in Texas, it was also something familiar and old, recalling an earlier era when aspersions were cast on Mexicans even as they were relied on to do difficult, unpleasant jobs. As the historian David Montejano documents in his book *Anglos and Mexicans in the Making of Texas*, in the 1920s and 1930s commercial farmers throughout Texas hired migrant workers from Mexico to toil in the fields. The migrants were appreciated for their work ethic and for

---

* On August 3, 2019, an assailant entered a Walmart in the border town of El Paso and murdered twenty-three people, the deadliest attack on Latinos in U.S. history. Law enforcement agents believe that the shooter, Patrick Crusius, posted a white nationalist manifesto on the message board 8chan beforehand, decrying the "Hispanic invasion" of Texas.

keeping wages down. ("Without the Mexican, the laboring class of white people . . . would demand their own wages and without doing half the labor the Mexican does," one grower remarked.) As neighbors and fellow citizens, they were appreciated far less. Most of the towns in rural Texas were strictly segregated, as were most schools. As Montejano notes, a recurrent theme in popular sentiment was that separation was necessary because Mexicans were "dirty," a term that connoted not just lack of hygiene but social untouchability. Among Anglos, "the Mexican was inferior, untouchable, detestable," a perception reinforced by the grubby jobs that migrant laborers performed, which came to be seen as "Mexican work." Mexicans "had to be taught and shown that they were dirty and that this was a permanent condition. They could not become clean."

Classifying Mexicans as dirty helped neutralize a perceived threat to the "social order," Montejano contends, ensuring that migrant workers would know their place even as they were depended on. As Montejano acknowledges, his analysis owed a debt to an influential work of cultural theory, the anthropologist Mary Douglas's *Purity and Danger*. Published in 1966, Douglas's book defined dirt as "matter out of place," something that came to be seen as repellent not because of its inherent foulness but because its existence could not be reconciled with the patterns and assumptions undergirding the existing social order. "Dirt offends against order," Douglas observed. For this reason, it was dangerous. Yet classifying something as dirty could also be affirming, because doing so implicitly marked off what was *not* dirty and needed to be kept clean. "Where there is dirt there is system," Douglas posited. "Dirt is the by-product of a systematic ordering and classification of matter, in so far as ordering involves rejecting inappropriate elements." The inappropriate element could be inanimate: filth, excrement. It could also be what Douglas called a "polluting person," a member of the community who has "crossed some line which should not have been crossed" and with whom others strained to avoid contact, not least to shore up their own standing as pure.

Whether the Mexican migrant workers in Montejano's study earned this designation because of what they did or because of who they were is unclear. In this respect, they were hardly unique. In India, a similar fate had long befallen *dalits*, or "untouchables," impoverished outcasts who were forced to perform defiling tasks like cleaning toilets and forbidden to have physical or social contact with upper-caste Indians. In Europe, no formal castes existed, but some ostracized groups still came to be seen as "polluting persons," among them Jews who served as moneylenders, an activity that was denounced as evil even as it became an increasingly ubiquitous and necessary feature of Western commercial life. There were plenty of Christian moneylenders in medieval Europe, but it was Jews who came to be regarded as the profession's most ruthless and cunning practitioners, greedy usurers accused of lending money to Christians at exorbitant rates. Church leaders condemned this as a mortal sin, even as many were quietly relieved to see the practice of lending money at interest (which the Bible forbade) outsourced to members of a rival faith whose salvation was not their responsibility. As Pope Nicholas V put it, it was preferable that "this people [Jews] should perpetrate usury than that Christians should engage in it with one another."

The repercussions for Jews were not entirely negative. As Simon Schama notes, moneylenders were the "potentates of England's Jews," sometimes living in lavish manors replete with elaborate fountains and hunting grounds. For the Jewish community as a whole, it was another matter. In 1518, a group of guilds in Regensburg, Germany, submitted a grievance charging that they and their fellow Christians had been "sucked dry, injured in their body and goods" by usurious Jews. One year later, more than five hundred Jews in Regensburg were summarily expelled. Arguably more damaging than any specific act of retaliation was the image of the "dirty Jew" that the association with moneylending instilled in the popular imagination, a stereotype that long outlasted the prohibition on charging interest. The machinations of "Jewish usurers" were described in *The Protocols of the Elders of Zion*, a canonical text in the literature of antisemitism. Several decades after the *Protocols* appeared, the official platform of the Nazi Party called for "the breaking of interest slavery," a practice that German antisemites, like their peers in other countries, associated with Jewish financiers. It scarcely mattered that as far back as the Middle Ages, Christians had frequently demanded higher interest rates on loans than Jews, much less that the earnings of Jewish moneylenders in places like England often wound up in the royal treasury, either through taxes imposed on the Jewish community or through confiscation upon death. "When a Jewish lender died, a third (at least) of their property reverted to the Crown, so that the hard bargains the Jews might have driven became a source of instant profit for the ever-voracious treasury," observes Schama. "The Jews were obliged to do the dirty work and get the odium while the Crown got the profit."

How would the "kill floors" in slaughterhouses continue to operate if fewer immigrants were around to do the dirty work? In August 2019, a revealing answer surfaced in Morton, Mississippi, after the Trump administration conducted a series of immigration raids on Mississippi's poultry plants. More than six hundred immigrants were arrested in the raids, which, in Morton, opened up jobs at a local chicken plant for members of Trump's political base, the white working class. Yet as *The New York Times* revealed in a story about the aftermath of the raids, few white Mississippians ended up applying for these jobs. Most of the applicants were African American, drawn by the fact that the plant paid $11.23 an hour, which was several dollars more than fast-food or retail jobs in the area paid. The extra income was appreciated. The accompanying moral complications were not. Many of the newly hired Black workers expressed misgivings about the raids, which they saw as racially motivated. "The way they came at the Hispanic race, they act like they're killing somebody," one worker told the *Times*.

In Mississippi as elsewhere, Black and brown workers were hired to do a job that many white people had evidently concluded was beneath them. In April 2020, a study found that whites made up just 19 percent of the meatpacking industry's frontline workers. People of color—Latino immigrants, African Americans, Asian immigrants from places like Vietnam and Myanmar—made up almost 80 percent. Nearly half of these workers lived in low-income households.

One of the low-income households I visited while in Bryan was Flor Martinez's residence, where she invited me for dinner one night. We sat in the kitchen of her two-story home, eating *sopa de nopal* (a Mexican stew flavored with cactus and tomatillo), while listening to the chirping of her pet parakeets, which lived in a white cage in an adjoining parlor. Over the course of the evening, Flor introduced me to her son and youngest daughter, both of whom still lived at home. She did not introduce me to Manuel, who no longer lived there. As Flor proceeded to explain, they'd gotten divorced, which meant she now had to make ends meet on her own, a challenge compounded by the fact that she no longer worked at Sanderson Farms. For the second—and, she assured me, final—time, she'd quit. She was now working at a Taco Bell in College Station, earning nine dollars an hour.

Flor did not hide the stress this had caused, mentioning a letter she'd just received from Wells Fargo, which informed her that if she didn't come up with eleven thousand dollars, she would soon be evicted from her house. Yet she did not express regret about her decision to stop working at Sanderson Farms. She simply *had* to stop, she said. When I asked her why, she reached for her iPhone, scrolled through some videos, and turned the device toward me. On the screen was a video she'd recorded of her right hand, which was curled into a ball and which she was massaging, methodically kneading the joints and knuckles with the fingers of her other hand. Every morning, she told me, she would go through this routine to try to alleviate the pain, which made it difficult for her to extend her fingers outward and turned simple tasks like getting dressed into forbidding challenges. "You see my finger—you see how it gets stuck," she said, pointing to the middle finger of her balled-up hand, which she kept trying to unfurl in the video, only to have it droop back down toward her palm, as though the tendon were a rubber band that had snapped. Sometimes, it took twenty minutes to coax her fingers open, she said. The indignity of the ordeal was compounded by the obligation she felt to document it. She'd made the video, she told me, after a nurse at the slaughterhouse told her she needed "proof" of her condition. Eventually, a doctor at the plant performed some tests and told her that the source of the pain was either lupus or arthritis, congenital problems that were not work related and therefore not the company's responsibility. "Your case is closed; it's been denied," a nurse informed her. Flor later went to see a doctor not connected to the plant who disputed this diagnosis, informing her that she did not have lupus or arthritis.

By this point, Flor said, "I was so angry. I hated the supervisor. I hated human resources. I hated everybody." Feeling hatred was not entirely new to Flor: as a teenager, she used to fantasize about killing the stepfather who tried to sexually abuse her. But the depth of her bitterness was new. The poverty of her upbringing, the drunken tirades of her grandfather, the experience of running away from an abusive home and making her way as an undocumented immigrant: none of this had managed to darken her sunny disposition. No matter how bleak things seemed, she had always believed that happiness was within her reach and that she would find a way to experience it. Working at Sanderson Farms shook this belief. As she felt it coming undone, she decided that keeping her job wasn't worth it—that, to preserve her self-respect, she had to leave.

Some time later, I learned that Flor had gotten a new job, at the Centro de Derechos Laborales, where she had volunteered on occasion and where the same qualities that had struck the owner of the Chick-fil-A years earlier—her diligence, her cheerful smile—drew the notice of her peers. The center was part of a coalition of labor and faith-based organizations that sought to educate and empower immigrants to fight back against degrading workplace conditions, both through workshops like the one I'd attended on sexual harassment and through direct actions like the demonstration that had followed it. At the demonstration, a group of protesters stood on a patch of grass by the entrance to Sanderson Farms, directly in front of the gate and private property sign, waving banners (EL BAÑO POR FAVOR) that decried the denial of bathroom breaks to workers on the lines. The protest received some coverage in the local news, which proved sufficiently embarrassing to Sanderson Farms that afterward the restrictions on bathroom breaks were eased. Some abusive supervisors were even fired.

Flor was still working at the center when, at the beginning of March 2020, she started hearing from workers who were concerned about a new hazard: the coronavirus. The workers had good reason to be concerned. Like prisons, the kill floors of America's slaughterhouses would soon be overrun with COVID-19, owing both to the crowded conditions in the plants, where workers often stood shoulder to shoulder, toiling in close quarters on the lines, and to the fact that at the outset of the pandemic the meatpacking industry devoted far more attention to continuing to operate at full capacity than to protecting its frontline employees. Sanderson Farms was no exception. Early on, Flor heard from frightened workers who complained that they were not being supplied with masks, much less spaced six feet apart from each other, in accordance with the social distancing guidelines issued by the Centers for Disease Control and Prevention. On March 20, 2020, Lampkin Butts, the company's president, sent workers a memorandum on "work attendance," informing them that as employees in a "critical infrastructure industry" they had a "special responsibility" to continue showing up for their shifts. The memo said nothing about slowing down the lines so that workers could stand farther apart from each other. Toward the end of March, workers were supplied with masks, and some hand-sanitizing stations were installed. The company also began requiring workers to have their temperature taken every morning, sending home anyone with a reading above a hundred degrees Fahrenheit. But those with other symptoms—runny noses, coughs—were not sent home, Flor said, and many felt the company was trying to hide the number of workers who had contracted the virus to avoid triggering panic.

One morning, as she was tracking all of this, Flor came down with a fever herself. She lay in bed for days, unable to move, her head pounding, her throat parched no matter how much water she drank. As she eventually learned, the cause was COVID-19. Flor battled the virus for days, fearing, at times, that she might not make it. "I was really, really sick," she said. Although the symptoms eventually abated, Flor had meanwhile learned that she had another potentially fatal disease: breast cancer. Flor relayed all of this to me in a string of text messages. In one of the messages, she mentioned that she was no longer working at the center, which, despite advocating for better working conditions for others, did not provide its own employees with health insurance or family medical leave, as she'd discovered while in quarantine. "I been fighting for the rights of workers that I thought I was given," she wrote. "I'm not planning to work for this place no more."

As dispiriting as all this was, Flor did not strike a note of self-pity or defeatism. She sounded optimistic, buoyed by the fact that, once again, she'd managed to overcome long odds. "I really think this is a miracle because I'm not supposed to be alive," she wrote, "and I'm here.

"I can't work now, so practically I have no income," she went on, "but we be ok. Mexicans are used to struggling and surviving."

7

"Essential Workers"

Dirty workers acted as "agents for the rest of us," Everett Hughes averred, carrying out unsavory tasks that many citizens tacitly condoned even as they distanced themselves from (and looked down on) the people who performed them. This was true of the guards who staffed the jails and prisons that served as America's de facto mental health asylums after states closed their psychiatric hospitals in the 1970s. It was true of drone operators who conducted targeted assassinations on behalf of an increasingly distracted and apathetic public after 9/11. In both of these cases—as in all of the examples that Hughes described in his essay—the people who did the dirty work were commissioned to perform their duties by the state. Their employer was the public, which underscored the fact that they were not rogue actors but agents of the body politic.

But in the United States, as in all modern societies, there is another way that citizens can support—and benefit from—dirty work that is allotted to others: not by serving as their employers, but by consuming the products they make. Flor Martinez worked for a private company, not the government, but the force that exerted the greatest pull over her industry was the appetite of the American people, consumers who ate prodigious quantities of chicken, beef, and pork while conveniently avoiding ever going near the production sites. One might think that health concerns and the growing popularity of once-exotic vegetables like kale have led Americans to scale back their consumption of animal protein. The statistics suggest otherwise. In 1960, the per capita consumption of meat and poultry was 161.8 pounds. Twenty years later, in 1980, it had risen to 192.9 pounds.

By 2000, the balance had shifted—less red meat, more chicken—but overall consumption continued to climb. After the 2008 Great Recession, the level fell slightly, most likely because families on tighter budgets could not afford as much meat, but the decline didn't last very long. In 2018, the Department of Agriculture forecast that the average American would eat 222.2 pounds of meat and poultry over the course of the year, a record figure and double the amount of animal protein that government nutritionists recommended. This was also more than twice the amount of meat that the average person on the planet ate.

What Americans wanted was cheap meat that could be consumed in abundance, the numbers suggested, which is one of the reasons that companies like Sanderson Farms and Tyson ran the lines in their slaughterhouses so fast, churning out as much product per dollar invested (and hour passed) as possible. This efficiency benefited everyone, the industry's spokespeople claimed, keeping meat affordable and enabling families to buy as much of it as they wanted. But as the author Michael Pollan has noted, the cheapness of American meat masks an array of hidden costs: to the environment (raising cattle is one of the largest sources of greenhouse gas emissions and a major cause of deforestation and water pollution); to public health (higher risks of heart disease and, thanks to the overuse of antibiotics, drug-resistant infections); to living animals, as even committed meat eaters would likely have acknowledged if shown what happened in America's factory farms and on the kill floors of industrial slaughterhouses. "No other country raises and slaughters its food animals quite as intensively or as brutally as we do," observed Pollan in his influential book *The Omnivore's Dilemma*. "No other people in history has lived at quite so great a remove from the animals they eat."

The remove is both physical and aesthetic. The steaks and drumsticks sold in supermarkets come in sterile, odorless packages that obscure the system's brutality. Some of the items that fill these packages—boneless patties, breaded nuggets—bear little resemblance to meat at all, making it easy to forget that an animal was killed to produce them. The desire to conceal this would not have surprised Norbert Elias, who, in *The Civilizing Process*, singled out the killing of animals as an example of how "disturbing events" came to be hidden in. In medieval societies, Elias noted, the upper classes routinely carved dead animals at the table, "not only whole fish and whole birds . . . but also whole rabbits, lambs, and quarters of veal." Over time, this norm gave way to "another standard by which reminders that the meat dish has something to do with the killing of an animal are avoided to the utmost. In many of our meat dishes the animal form is so concealed and changed by the art of its preparation and carving that while eating one is scarcely reminded of its origin."

Elias went on: "Specialists take care of it in the shop or the kitchen . . . The curve running from the carving of a large part of the animal or even the whole animal at table, through the advance in the threshold of repugnance at the sight of dead animals, to the removal of carving to specialized enclaves behind the scenes is a typical civilization-curve."

To eat meat in America in the twenty-first century was to stand at the far end of this curve, at a safe remove from the repugnant sights that slaughterhouse workers encountered on a daily basis and that consumers were not even permitted to glimpse on television. In her book *Slaughterhouse*, the animal rights activist Gail Eisnitz described her attempts to persuade a senior producer at *20/20* to air a segment on the mistreatment of livestock that she'd uncovered at several meat and poultry plants. The producer was intrigued, but the idea was ultimately rejected, out of concern that the material would be "too graphic for the viewing public." The night Eisnitz learned this, she surfed the channels on her television, taking in a show about a cop who beat a confession out of a prisoner, a drama in which a rape occurred, and the nightly news, which consisted of "a solid half hour of war, starvation, and genocide," none of which was considered too graphic for the viewing public.

The political scientist Timothy Pachirat has described the industrial slaughterhouse as a "zone of confinement," an isolated, violent place "inaccessible to ordinary members of society" (Pachirat borrowed the term from the sociologist Zygmunt Bauman). To penetrate this zone, Pachirat decided to apply for a job at a beef slaughterhouse, a plant whose workforce consisted mainly of immigrants and refugees and whose kill toll exceeded ten thousand cows a week—twenty-two hundred per day, a figure that gave rise to the title of a book he later published about his experience, *Every Twelve Seconds*. To kill so many animals at such a pace was messy, Pachirat discovered, splattering the workers in manure, blood, vomit, and offal (the cow's entrails and internal organs). It also required a good deal of brutality and violence. At one point during the five and a half months he worked there, Pachirat was assigned to the chutes, where workers used electric prods to maneuver cows through a serpentine corral that led to the slaughterhouse. "The cattle jump and kick when shocked . . . and many also bellow sharply," he wrote. When a cow collapsed in one of the chutes, the line kept moving, causing the downed animal to get stomped on. From the chutes the cows proceeded along a conveyer to the knocking box, where they were shot in the head with a captive-bolt gun. One day, Pachirat entered the box and was shown how to use it. After aiming the bolt a few inches above the cow's eyes, he pulled the trigger and watched blood spurt out of its skull. The slain animal was carted off on the conveyer; within seconds, another appeared in the knocking box, "head swinging and eyes large in terror." A coworker later warned Pachirat to avoid working as a knocker. "That shit will fuck you up," he said. The warning resonated, not least because, during his time in the chutes, Pachirat repeatedly came into direct contact with the cows, patting their noses and gazing at their velvety hides. "Fucked up is exactly how I feel," he wrote afterward. "It is how I would describe many of the chute workers, and it captures the rawness and violence of the perpetual confrontation between the living animals and the men driving them."

"The worst thing, worse than the physical danger, is the emotional toll," a worker at a pork slaughterhouse told Gail Eisnitz. "Pigs down on the kill floor have come up and nuzzled me like a puppy. Two minutes later I had to kill them." Many of his coworkers abused drugs or alcohol to try to numb themselves, the worker said. "Only problem is, even if you try to drink those feelings away, they're still there when you sober up." The statement reminded me of the stories I'd heard from prison guards like Tom Beneze, who worked in a profession that drove people to rely on similar coping mechanisms. Killing animals for a living could have an equally profound effect on workers' psyches, Eisnitz's account suggested.

Pachirat came to a different conclusion. Working as a knocker could indeed cause severe emotional distress. "Nobody wants to do that," another worker told him. "You'll have bad dreams." But most of the workers at the plant did not seem troubled by what they were doing. A major reason for this was the division of labor on the kill floor, which was segmented into so many different subspecialties—belly ripper, liver hanger, head chiseler—that the act of killing became fragmented (only the "knocker" actually killed the animal, many workers were convinced). This description matched the impression I gleaned from the one kill floor that I managed to see firsthand, at a poultry slaughterhouse where workers were given similarly specialized roles: one group harvesting livers, another slicing necks, a third trimming "defects." As the workers pruned and gutted the birds, rivulets of pink bloodied wastewater ran along the floor, sloshing onto the soles of my shoes. It was unsettling, and the smell on the kill floor caused me to gag at first. Yet it was easy to imagine how, after a while, slicing necks or harvesting livers became mundane, a technical task devoid of anguish, which is how the Sanderson Farms workers I'd interviewed described their jobs to me and how, over time, even the chute workers at the slaughterhouse where Pachirat got a job came to feel, mocking him for being soft. "You motherfucking pussy!" one worker shouted when Pachirat refused to use an electric prod to push the cows toward the knocking box. "What's the point of shocking them?" Pachirat asked. "The point is pain and torture," the worker said.

To judge by this exchange, organizations like PETA are not wrong that working in a slaughterhouse can instill a blasé attitude toward suffering. It can foster cruelty and sadism. But who bears more responsibility for this cruelty: the workers who shock and kill the animals (and whom some PETA activists have advocated charging with criminal felonies), or the consumers who eat meat without ever thinking about the costs? After quitting his job at the beef slaughterhouse, Pachirat debated this question with a friend. "She maintained, passionately and with conviction, that the people who did the killing were more responsible because they were the ones performing the physical actions that took the animals' lives," he wrote. Pachirat argued otherwise, insisting to his friend "that those who benefited at a distance, delegating this terrible work to others while disclaiming responsibility for it, bore more moral responsibility, particularly in contexts like the slaughterhouse, where those with the fewest opportunities in society performed the dirty work."

"FULFILLING ORDERS"

There is no way for this work to be anything but terrible, some would argue. One could hardly begrudge animal rights activists from organizations like PETA for feeling this way. Even organizations that do not share PETA's agenda know better than to portray the slaughter of livestock for mass consumption as a pleasant undertaking. "Killing and cutting up the animals we eat has always been bloody, hard and dangerous work," noted Human Rights Watch in its 2005 report on labor conditions in the meatpacking industry. At the turn of the twentieth century, the stockyards where the killing was done "were more than sweatshops," the report went on. "They were blood shops, and not only for animal slaughter. The industry operated with low wages, long hours, brutal treatment, and sometimes deadly exploitation of mostly immigrant workers."

And yet, as with other forms of dirty work, the conditions in the meatpacking industry were not foreordained. They were shaped by rules and regulations and by government agencies that, in theory at least, had the power to make the work less terrible, both with respect to how the animals were treated and with respect to how the workers were. Whether agencies would actually do so was another matter, of course. Following passage of the 1906 Meat Inspection Act, which was signed into law shortly after Upton Sinclair published *The Jungle*, the government began to regulate the meatpacking industry to prevent a recurrence of the unsanitary practices that had shocked and horrified Sinclair's readers. In the decades after World War II, the U.S. Department of Agriculture dispatched federal inspectors to slaughterhouses to examine carcasses and, if necessary, halt production in order to remove contaminated meat from the lines. The system wasn't perfect, but it functioned well enough to assure consumers that the meat they put on their dinner plates was safe to eat. In the 1980s, however, the Reagan administration embraced a new system of "streamlined inspection." Touted as more modern and scientific, streamlined inspection enabled companies to accelerate production while cutting the number of federal inspectors on the lines. In 1991, the reporter Scott Bronstein published an exposé in *The Atlanta Journal-Constitution* that described the consequences in the poultry industry. "Every week throughout the South," Bronstein reported, "millions of chickens leaking yellow pus, stained by green feces, contaminated by harmful bacteria, or marred by lung and heart infections, cancerous tumors or skin conditions are shipped for sale to consumers, instead of being condemned and destroyed." Bronstein quoted inspectors who told him that the USDA seal of approval ensuring that poultry was safe for human consumption had become "meaningless." "We're well aware of the problem," one industry spokesman acknowledged. "And we don't have an answer for it yet."

Six years later, in 1997, the Clinton administration rolled out an industry-friendly solution to the problem—the so-called Hazard Analysis and Critical Control Point system. Framed as a watershed reform that would improve safety, the new system instead transferred authority to quality assurance officers employed by the industry and reduced federal inspectors to secondary spot-check duty. Some federal inspectors jokingly called HACCP "Have a Cup of Coffee and Pray." An updated version of HACCP—known by the acronym HIMP—eventually emerged and was given a different label: "Hands in My Pocket." As the number of federal inspectors declined, companies took to spraying meat with bacteria-killing chemicals such as peracetic acid, or PAA, and chlorine. The sprays—which were banned in Europe, where consumers overwhelmingly opposed consuming meat doused in chemicals—did nothing to address the filthy conditions on factory farms that exposed so many animals to disease. They did nothing to halt excessive line speeds. It was not even clear that they were effective at killing off dangerous pathogens. The one clear advantage the chemicals did have was that they were cheap, enabling the USDA to further reduce staffing and allowing companies to speed production even more. As one USDA poultry inspector told *The Washington Post*, "They don't talk about it publicly, but the line speeds are so fast, they are not spotting contamination, like fecal matter, as the birds pass by . . . Their attitude is, let the chemicals do the work."

Spraying chemicals inside industrial slaughterhouses had another, less noticed effect, which was to sicken many of the workers who inhaled them. Poultry plants were required to post Material Safety Data Sheets to warn workers about the potential health risks associated with the use of these chemicals. For PAA, these risks included "damage [to] most internal organs, including the heart, lungs and liver." In 2011, an inspector at a poultry plant in New York State where PAA and chlorine were used died after his lungs bled out. In the years that followed, complaints about the air quality in poultry plants surfaced with growing frequency. At the workshop I attended in Bryan, Texas, I heard such complaints again and again—from poultry workers in Arkansas, from their peers in North Carolina, from Flor Martinez, who told me that her throat and lungs were constantly irritated when she worked at Sanderson Farms. "Your throat—it feels like you have this burning," she said. Yet neither the USDA nor the Food and Drug Administration had reviewed whether exposing workers and its own inspectors to the chemicals was safe. Nor had the government set any permissible exposure limits for PAA.

By the spring of 2020, the workers in America's industrial slaughter-houses had reason to fear breathing in something else: respiratory droplets laced with the novel coronavirus that causes the potentially deadly COVID-19. Like prison guards, the people who toiled on the kill floors of these slaughterhouses were designated "essential workers" during the pandemic and instructed to continue doing their jobs. And like prison guards, many discovered that this designation did not entitle them to personal protective equipment or safe working conditions, much less to the kind of public recognition bestowed on medical workers and first responders. According to the Centers for Disease Control and Prevention, more than sixteen thousand slaughterhouse workers had tested positive for COVID-19 and eighty-six workers had died by the beginning of July. The actual toll was likely far greater, both because of delays in identifying outbreaks and because only twenty-one states had submitted data to the CDC. Though incomplete, the CDC data was nevertheless revealing, showing which communities were absorbing the devastation and which were not. Of the meatpacking workers who had died of COVID-19, 87 percent were racial or ethnic minorities.

One of the states that did submit data to the CDC was Colorado, which was home to a beef slaughterhouse in Greeley where, in April, an outbreak of COVID-19 caused a wave of infections and three worker deaths. After a county analysis showed that 64 percent of employees who'd tested positive had continued reporting to work—reflecting what one local health official called a "work while sick" culture at the plant—Colorado ordered the slaughterhouse to close. Yet eight days later, the facility reopened and the outbreak continued to spread. Soon, three more workers were dead, all of them minorities. As *The Washington Post* revealed in a story published a few months later, in the intervening period, JBS, the company that owned the plant, had enlisted a powerful ally—the White House. Soon thereafter, Robert Redfield, the director of the CDC, called Jill Hunsaker Ryan, the head of Colorado's health agency, apparently at the behest of Vice President Mike Pence. "JBS was in touch with the VP who had Director Redfield call me," Ryan wrote in an email to a county health official. According to the email, Redfield wanted the state to permit the company to send "asymptomatic people back to work even if we suspect exposure but they have no symptoms." In its story, the *Post* traced the effects on the family of Bienvenue Chengangu, a Congolese refugee who began to cough and develop a fever after completing a night shift. Chengangu soon tested positive for the virus and ended up infecting his mother, who was seventy-three and had high blood pressure. Although Chengangu recovered, his mother did not, leaving him to wonder whether his decision to flee the civil strife in the Congo for the apparent safety of the United States had been wise. "I'm realizing America isn't the paradise we believed it to be," he said.

The JBS plant in Greeley was not the only slaughterhouse that the federal government pressured state officials to keep open. On April 26, 2020, a full-page ad about the dire situation in the meatpacking industry appeared in *The New York Times*—dire not for the workers dying and falling ill but for consumers who might be deprived of meat. "The food supply chain is breaking," proclaimed the ad, an open letter from John Tyson, the CEO of Tyson Foods. "As pork, beef and chicken plants are being forced to close, even for short periods of time, millions of pounds of meat will disappear from the supply chain," the ad warned. "As a result, there will be limited supply of our products available in grocery stores until we are able to reopen our facilities."

Two days after the Tyson ad appeared, President Donald Trump invoked the Defense Production Act to order meatpacking plants to remain open, overruling states and localities that had temporarily shuttered them for health reasons. "It is important that processors of beef, pork, and poultry ('meat and poultry') in the food supply chain continue operating and fulfilling orders to ensure a continued supply of protein for Americans," the executive order affirmed. As press accounts soon revealed, supplying protein for Americans was not the meatpacking industry's top priority. In the same month that its full-page ad appeared in the *Times*, Tyson shipped 2.5 million pounds of pork to consumers in another country—China. Smithfield, whose CEO also raised the specter of meat supply shortages in U.S. grocery stores, exported 18 million pounds of pork to China. "In all, a record amount of the pork produced in the United States—129,000 tons—was exported to China in April," noted an article in the *Times*.

The revelations did not stop the Trump administration from helping the meatpacking industry with another matter it saw as urgent: protection from lawsuits filed by workers forced to toil in crowded, unsanitary plants overrun with COVID-19. As long as companies made "good faith attempts" to comply with the CDC's health and safety guidelines, they would not be held legally responsible for exposing workers to the virus, a memo issued by the administration affirmed. The memo came from the Department of Labor and was cosigned by Loren Sweatt, the principal deputy assistant secretary for the Occupational Safety and Health Administration. OSHA was founded in 1970, with the mission "to provide a safe and healthy workplace for every working man and woman in the Nation." Fulfilling this mandate had never been easy, both because of limited resources—OSHA's budget is roughly one-tenth the size of the Environmental Protection Agency's—and because of strong opposition from the business community, which invariably depicted the agency's rules as an unwarranted intrusion on the free enterprise system. Even so, when the pandemic struck, it stood to reason that OSHA would assume a more prominent role. One thing the agency could have done was to impose an emergency temporary standard requiring companies to follow specific rules to protect workers from COVID-19. The template for such a directive already existed, thanks to the 2009 H1N1 pandemic, which prompted the Obama administration to launch an initiative to develop an infectious disease standard for work sites. As the death toll from COVID-19 climbed and meatpacking plants across the country reported outbreaks, some staffers at OSHA prepared a version of the standard that they hoped would be rolled out.

No emergency standard was issued. Instead, OSHA put out a series of memorandums listing guidelines that various industries could follow—guidelines that were not binding and that companies were explicitly told created no new legal obligations. The lax approach led some union leaders and worker-safety advocates to accuse OSHA of shirking its mission. It did not surprise anyone familiar with the background and beliefs of Eugene Scalia, who was secretary of labor at the time. The son of the late Supreme Court justice Antonin Scalia and a partner at the white-shoe firm Gibson, Dunn & Crutcher, Scalia did not come to the Department of Labor with a reputation as a stalwart advocate of workers' rights. He did come to it with intimate knowledge of government rules and regulations, including worker protections enforced by OSHA—regulations he'd spent much of his career trying to weaken and undermine.

Scalia first rose to national prominence in the spring of 2000, at a hearing that the Department of Labor organized to solicit input from the public on a draft version of a new ergonomics standard that OSHA had been developing. The standard was designed to address musculoskeletal disorders such as carpal tunnel syndrome and tendinitis, injuries that afflicted hundreds of thousands of workers every year, including an untold number who worked in poultry slaughterhouses like the one I'd visited in Bryan. Labor unions had been pushing for such a standard for years. But business lobbyists opposed the idea. At the hearing, which was held in a windowless auditorium on the ground floor of the Frances Perkins Building in Washington, D.C., Scalia did their bidding, grilling witnesses who came forward to testify in support of OSHA's draft rule and questioning the research linking musculoskeletal disorders and workplace risk factors. The evidence supporting such a link was voluminous, including a National Institute for Occupational Safety and Health review of more than six hundred studies that found "a consistent relationship between MSDs and certain physical factors, especially at higher exposure levels." Scalia was unimpressed, portraying ergonomics as 'junk science' both at the hearing and in a report for the Cato Institute, which argued that "supposed musculoskeletal disorders" correlated more with "psychosocial factors" such as whether workers liked their jobs than with occupational risk factors.

Eric Frumin, who was then the health and safety director for the Union of Needletrades, Industrial, and Textile Employees, also attended the OSHA hearings. Even as he dismissed ergonomics as "junk science," noted Frumin, Scalia was representing companies such as UPS that had adopted ergonomic principles to prevent their own workers from sustaining injuries. Frumin likened the strategy to oil companies that publicly denied the scientific evidence of climate change even as they privately accepted its validity. "The level of deceit is every bit as vicious as what we've seen on climate change, or what Purdue said about Oxycontin," he said. "It doesn't get the same level of attention, because it's about workers who break a sweat every day. But it's every bit as dangerous in blocking the use of science to protect people from severe preventable risks to their health."

As it turned out, Scalia's efforts did not succeed in blocking OSHA from adopting an ergonomics standard, which it announced in November 2000. The regulation would prevent as many as four million workers from sustaining occupational injuries over the next decade, the agency predicted, sparing untold numbers of workers in warehouses and poultry slaughterhouses from suffering. But the rule didn't last long. After George W. Bush became president in 2001, Republicans invoked a provision of Newt Gingrich's "Contract with America" that allowed Congress to conduct expedited reviews of government regulations and voted to overturn it. Without an ergonomics standard on the books, workers exposed to repetitive strain injuries were left to rely on OSHA's General Duty Clause, which required employers to create an environment "free from recognized hazards." The statute's vague language made enforcement difficult and time-consuming, raising the burden of proof for corroborating violations. In 2002, United Food and Commercial Workers International submitted a complaint to OSHA about a poultry slaughterhouse in Lufkin, Texas, owned by Pilgrim's Pride that it thought might meet the burden. "I haven't seen conditions like this in 20 years," Jackie Nowell, director of the UFCW's safety and health office, said after visiting Lufkin and talking to workers at the plant, many of whom were undocumented immigrants too frightened to say anything to management. The UFCW filed the complaint in the hope that even in the absence of an ergonomic standard employers would be held accountable for egregious violations. But the conditions that shocked Nowell were not enough to move OSHA, which dismissed the complaint.

"COMMERCIAL SHACKLES"

Had the ergonomics standard remained on the books, the lives of the workers at the Pilgrim's Pride slaughterhouse in Lufkin, like the lives of Flor Martinez and the other workers I met in Bryan, might have been different. They might have not only sustained fewer physical injuries but also suffered fewer indignities, the blows to the spirit that Angela Stuesse described. Politicians and public officials could have done plenty of other things to lessen these blows, from strengthening the penalties that OSHA could impose on companies that willfully violated health and safety standards to slowing the line speeds in slaughterhouses.

What stopped these things from happening were not immutable structural conditions but political forces that mobilized to prevent them and real people driven by specific ideas and agendas. In a different country or simply a different era, one of the forces that might have mobilized to push for slower line speeds and better working conditions in America's slaughterhouses was organized labor. But as in so many other sectors of the U.S. economy, the power of unions had radically declined in the meatpacking industry since the 1970s. Unlike the master agreements negotiated by the United Packinghouse Workers after World War II, unions often had little control over wages and working conditions even in plants where they were still around. Their clout was especially limited in the poultry industry, as was apparent at the Sanderson Farms plant in Bryan, which actually had a union—a union that was unable to prevent management from subjecting workers to degrading treatment (none of the workers I interviewed belonged to it).

While the power of unions had diminished, the influence of another force—organized money—had grown. OSHA's ergonomics standard might well have survived if not for the National Coalition on Ergonomics, an industry lobby created by the National Association of Manufacturers and the U.S. Chamber of Commerce that spent years portraying such a rule as a threat to American competitiveness. One of the lawyers representing the lobby was Eugene Scalia. Among the ideas that drove Scalia was the conviction that government regulation of the private sector was harmful and illegitimate. Since the Reagan era, this belief had grown increasingly influential and widespread, propagated in reports put out by conservative think tanks such as the Cato Institute and nurtured at institutions such as the University of Chicago, where Scalia attended law school. (The school was home to the influential law and economics movement, which applied the principles of neoclassical economics to legal reasoning, an approach championed by many of the conservative judges appointed to the federal bench by the Trump administration.) Critics of regulation were sometimes fond of quoting James Madison's observation that "commercial shackles are unjust, oppressive and impolitic" and that industry and labor should therefore be left alone, free from interference by "enlightened legislatures." As his performance at the ergonomics hearing in 2000 attested, Scalia was a dedicated proponent of this view. A decade later, he trained his attention on another set of "commercial shackles," the regulations codified in the Dodd-Frank Act, which Congress enacted in 2010 to rein in the reckless conduct that precipitated the 2008 financial crash. After the law went into effect, Wall Street hired a phalanx of high-powered lawyers to challenge its provisions in court, a role Scalia took up with such alacrity that some staffers at the Securities and Exchange Commission jokingly referred to Dodd-Frank as the "Eugene Scalia Full Employment Act."

Scalia's long record of opposing worker protections and representing financial interests explains why many union leaders viewed his nomination to serve as secretary of labor as an affront. It also explains why he was chosen for the role by Donald Trump, who campaigned in 2016 as a champion of ordinary workers but who, upon entering office, quickly made it clear that he had other priorities: cutting taxes for the rich, shredding environmental and labor protections. At the start of his term, Trump announced that all federal agencies had to revoke at least two regulations for every new one that was added. Under Alex Acosta, who served as secretary of labor during Trump's first two and a half years in office, the Department of Labor did eliminate some regulations. But hardliners within the Trump administration were dissatisfied with the pace of change. Acosta was eventually replaced by Scalia, who accelerated the pace, in particular by eliminating a series of little-known rules that protected low-wage workers from wage theft and other exploitative practices. It was an odd agenda for the Department of Labor, whose official role is "to foster, promote and develop the welfare of the wage earners, job seekers and retirees of the United States." Yet it was consistent with Scalia's other actions, including some new rules he quietly introduced that benefited more powerful actors. Among these rules was a proposal making it easier for companies to classify millions of gig workers as "independent contractors" (unlike regular employees, such contractors can be denied the minimum wage, overtime pay, and other benefits). As this suggested, Scalia's opposition to regulations was selective, masking what appeared to be the real agenda, which was to augment corporate power—in the case of the independent contractors rule, the power of companies like Uber and DoorDash, which happened to be clients of Gibson Dunn, the law firm where he'd worked before joining the Trump administration.

As when Congress repealed the ergonomics standard two decades earlier, organized money was able to make its voice heard. As in the past, organized labor was not. On April 28, 2020, Richard Trumka, the head of the AFL-CIO, sent Scalia a letter urging the Department of Labor to issue emergency temporary standards to protect "essential workers" in meatpacking and other sectors of the economy from COVID-19. In response, Scalia insisted that doing so was unnecessary because OSHA already had the authority to penalize irresponsible companies, under the General Duty Clause. By the end of October 2020, OSHA had received more than ten thousand complaints alleging unsafe conditions related to the virus. It had issued just two citations under the General Duty Clause. The agency eventually issued a few more citations, including to two meatpacking plants. One of the plants, a pork slaughterhouse owned by Smithfield Foods in Sioux Falls, South Dakota, where twelve hundred workers fell ill and four died, was fined $13,494. The other was the beef slaughterhouse in Greeley, Colorado, where a "work while sick" culture had prevailed. It was fined $15,615. Deborah Berkowitz, who served as OSHA's chief of staff under President Obama, described the penalties as "a slap on the wrist" to two billion-dollar corporations that, far from deterring other companies, would embolden them. "The Department of Labor's job is to ensure that businesses that cut corners on safety will know they'll face serious consequences if they endanger their workers," said Berkowitz. "This paltry fine sends the opposite message, telling companies, 'Don't worry if workers get sick or die on the job.'" Like many observers, Berkowitz was convinced the lenient treatment was not unrelated to the fact that so many slaughterhouse workers were immigrants and people of color. "These are Black and brown workers, and I just don't think this administration cares about them at all," she said. Before the pandemic, the agency under Trump that paid the closest attention to what transpired in industrial slaughterhouses wasn't OSHA. It was ICE, which, in 2018, dispatched agents to round up ninety-seven immigrants at a meatpacking plant in Bean Station, Tennessee, the largest workplace raid in a decade. During the pandemic, the raids appeared to cease, perhaps out of recognition that immigrants were actually needed to keep the lines in slaughterhouses running. Notably, however, no effort was made to call attention to the sacrifices these immigrants made to help keep the nation fed. In August 2020, Trump appeared in a video with an assortment of "essential workers"—a postal worker, some nurses, a truck driver. "These are great, great people," he proclaimed. Filmed in the East Room of the White House, the segment aired on national television, during the Republican National Convention. No slaughterhouse workers appeared in it.

"REGULATION BY SHAMING"

To be sure, America is not the only country that relies on foreign laborers to work in slaughterhouses. In Germany, the dirty work in these facilities has often been delegated to transplants from poorer neighboring states (Bulgaria, Romania) who are hired through subcontractors, housed in crowded quarters, and ruthlessly exploited. Perhaps not surprisingly, outbreaks of COVID-19 erupted in some German slaughterhouses as well. In June 2020, fifteen hundred workers at a plant in North Rhine–Westphalia tested positive for the virus, prompting the authorities to close local schools and order a lockdown. But when Tönnies, the company that owned the facility, tried to blame the outbreak on the plant's foreign workers, its owners were immediately rebuked. Hubertus Heil, Germany's labor minister, accused the company of taking "an entire region hostage" and demanded that it pay damages. After Tönnies apologized and offered to pay for widespread testing in the community, Heil was unappeased, telling the press he had "zero" trust in the company. He later called the meatpacking industry a system of "organized irresponsibility" and proposed "fundamental" change, including increased monitoring and a ban on subcontractors.

Here was another way that public officials could respond when dangerous and degrading work conditions came to light: by shaming the companies that profited from them. This approach was not unheard of in America. In 2009, OSHA launched an initiative to shine a light on companies that willfully violated the law. In 2014, after four workers at a DuPont facility in Texas were asphyxiated following exposure to a toxic chemical, David Michaels, an epidemiologist who headed OSHA at the time, declared, "*Nothing* can bring these workers back to their loved ones . . . We here at OSHA want DuPont and the chemical industry as a whole to hear this message loud and clear." According to Matthew Johnson, an economist at Duke and the author of "Regulation by Shaming," a study of the policy's deterrent effects, targeting such messages at local media and industry trade publications led to a 30 percent reduction in violations at adjacent facilities in the same industry. A single news release could have the impact of two hundred OSHA safety inspections, the study found.

The Trump administration took a different approach, one outlined in a September 24, 2020, memorandum, in which Patrick Pizzella, the deputy secretary of labor, instructed OSHA and other enforcement agencies not to publicize company violations "absent extraordinary circumstances." Press releases about COVID-related violations promptly stopped. Protecting the reputation of meatpacking companies was evidently more important to the government than protecting the lives of slaughterhouse workers. The low value placed on the lives of these workers came to light when a nonprofit organization called Justice at Work sued OSHA for failing to protect meat packers at a Maid-Rite plant in Pennsylvania. In March 2020, OSHA had received a complaint alleging that the conditions at the plant, where workers toiled in cramped quarters with no barriers between them, posed an "imminent danger." The agency, without bothering to inspect the facility, determined that the danger was not imminent. When an inspection finally was approved several months later, in July, OSHA contacted the plant's director of human resources beforehand to announce that it was coming. At a hearing, Matthew Morgan, a lawyer for

Justice at Work, asked if giving a company a heads-up was conventional practice. It wasn't, an OSHA inspector indicated. "Then why did you do it here?" Morgan asked. "To make sure that I was safe from COVID-19," the inspector said, explaining that a "job-hazard analysis" had been done on the *inspectors'* behalf and that her superiors had determined that taking added precautions was justified before entering the facility. "OSHA has a right to protect employees," the inspector told Morgan—a right that Maid-Rite workers could evidently do without.

Along with lax oversight came something else: waivers from the USDA granting fifteen poultry slaughterhouses permission to run their lines even faster. According to the National Employment Law Project, all of the plants that received the waivers had records of severe injuries, had been cited for violations by OSHA in the past, or had COVID-19 outbreaks at the time.

The coronavirus pandemic did occasion some shaming in America as well—not of the meatpacking industry, as in Germany, but of the workers themselves. In a call with a bipartisan group of congressional representatives, Alex Azar, the secretary of health and human services, suggested that the outbreaks of COVID-19 in meatpacking plants had more to do with the "home and social" aspects of workers' lives than with workplace conditions. Kristi Noem, the Republican governor of South Dakota, asserted that "99%" of the cases at meatpacking plants in her state could be attributed to the fact that the workers lived "in the same community, the same buildings, sometimes in the same apartments." As with guards who were blamed for having a penchant for cruelty when abuses in prison came to light, such statements implied that the key moral failure rested with a few reckless individuals—"bad apples"—rather than with the exploitative system in which they worked.

Such statements touched a nerve in Dulce Castañeda, a resident of Crete, Nebraska, which was home to a pork slaughterhouse owned by Smithfield. Castañeda did not work at the plant. Her father, a Mexican immigrant, did. At the start of the pandemic, Castañeda learned that her father and other workers at the slaughterhouse were given hairnets rather than N95 or surgical masks. At one point, the company announced that it was closing the plant, but then abruptly reversed the decision and kept it open. Although Smithfield eventually made masks available and installed plastic barriers between workers on the disassembly line, these barriers were removable, Castañeda's father told her, and many workers still believed they were at high risk. Castañeda's father had underlying health conditions, yet, like many workers at the plant, he was too afraid to submit a complaint to OSHA, fearing it would trigger retaliation from the company. On May 27, 2020, Castañeda decided to file a complaint of her own. In it, she listed an array of concerns, among them that social distancing rules were not enforced in the cafeteria or bathrooms and that employees were not kept informed about which of their coworkers were exposed, positive, or sick. By this point, more than one hundred workers at the plant had fallen ill, helping to explain why Saline County, where Crete was located, had one of the highest per capita rates of infection in the country.

After contacting OSHA, Castañeda was cautiously optimistic. "I had a little bit of hope that perhaps if they start receiving these complaints, then maybe that would encourage them to get out here and actually see for themselves what's happening," she said. Her hopes were buoyed when an OSHA officer called her immediately. The officer proceeded to explain that maintaining social distance was the responsibility of the workers, not the company. "He said, 'The employer cannot sit there and watch them all day,'" she recalled. When Castañeda asked whether OSHA might conduct an on-site inspection, the officer said this was unlikely because Smithfield was "complying with everything."

At the end of the call, the OSHA officer said one other thing, recommending that Castañeda tell her family members to wear masks and to practice social distancing. The advice was "a slap in the face," she said, as if she and her family members had not *already* taken these steps and as though the problem were not the conditions at the plant but their personal behavior. Her bitterness deepened when, a few weeks later, she came across an article quoting a letter that Kenneth Sullivan, the CEO of Smithfield, sent to Nebraska's governor, Pete Ricketts, in which he described social distancing as "a nicety that makes sense only for people with laptops." Ricketts could have responded to this insulting claim the way Hubertus Heil did to Tönnies in Germany. That is, he could have shamed the company for holding places like Saline County hostage and announced that he would do everything in his power to reform the industry. Instead, he fixed the blame on a more convenient target: the workers. "What do we see in places where we see a lot of spread of the virus?" said Ricketts. "Well, we see people concentrate together, and that's certainly true with our food processors." Soon, meatpacking workers in Nebraska found themselves being turned away from retail stores or asked where they worked before getting haircuts, Tony Vargas, a state senator from Nebraska, told NPR. In other words, they started to be treated like "polluting persons" whose own moral failings were the problem, which is how dirty workers are often made to feel. This was how Dulce Castañeda felt after coming across a story in which another Smithfield spokesperson blamed the outbreak of COVID-19 on "living circumstances in certain cultures," which "are different than they are with your traditional American family."

"That's saying that we don't live in good conditions," she said with dismay. "That we're dirty."

VIRTUOUS CONSUMERS

While "essential workers" in the poultry industry were made to feel dirty, nonessential workers in fields like finance and computer engineering—the "people with laptops"—were sheltering in place, more distant from what transpired in industrial slaughterhouses than ever before. Thanks to FreshDirect and Instacart, consuming meat no longer even requires coming into contact with a deli butcher or grocery clerk. With a few taps on a keyboard or the swipe of a screen, consumers can get as much beef, pork, and chicken as they want delivered to their doors, without ever having to think about where it comes from.

And yet, as the popularity of bestselling books like Michael Pollan's *The Omnivore's Dilemma* and Jonathan Safran Foer's *Eating Animals* attests, a lot of Americans do think about this. In recent years, more and more consumers have begun to carefully scrutinize the labels on the packages of the meat and poultry they buy. The ranks of such consumers have grown exponentially, paralleling the rise of the "good food" movement, which promotes healthier eating habits and reform of the industrial food system. Although the movement is, in Pollan's words, a "big, lumpy tent," composed of a broad coalition of advocacy organizations and citizens' groups that sometimes push for competing agendas, one of its aims is to persuade consumers to become more conscientious shoppers and eaters. Among those who put this idea into practice are so-called locavores, who buy food directly from local farms, ideally from small family-run enterprises that embrace organic, sustainable practices: ranchers who raise grass-fed cows that never set foot in industrial feedlots; farmers who sell eggs that come from free-range chickens reared on a diet of seeds, plants, and insects rather than genetically engineered corn and antibiotics.

Locavores engage in what social scientists call "virtuous consumption," using their purchasing power to buy food that aligns with their values. The movement appeals to the growing number of Americans who want to feel more connected to the food they eat and to the people who raise it, with whom locavores can interact directly at farmers markets or through community-supported agriculture programs. It is a captivating vision, and the benefits of eating locally grown food—which is likely to be more nutritious, to come from more humanely treated animals, and to be better for the environment—are manifold.

But locavores have some blind spots of their own, most notably when it comes to the experiences of workers on small family farms. As the political scientist Margaret Gray discovered when she set about interviewing farm laborers in New York's Hudson Valley, the vast majority of these workers are undocumented immigrants or guest workers who toil under abysmal conditions, often working sixty- to seventy-hour weeks for dismal pay. "We live in the shadows," one worker told her. "They treat us like nothing," said another. In her book *Labor and the Locavore*, Gray asked the butcher on a small farm why so few of his customers seemed to notice this.

"They don't eat the workers," the farmer told her.

"He went on to explain that, in his experience, his consumers' primary concern is with what they put in their bodies," Gray wrote, "and so the labor standards of farmworkers simply do not register as a priority."

The farmer's observation helps clarify why ethical eaters often seem far more attuned to the welfare of the animals in the food system, taking pains to buy beef and chicken labeled "cage free" and "Certified Humane," than to the welfare of the workers, about whom the labels say nothing. It also underscores the limits of virtuous consumption, which can reduce politics to a market transaction whose main purpose is to make individuals feel better about themselves. These individuals often care more about their own health and about a certain kind of purity—the purity of not eating meat laced with antibiotics, the purity of keeping artificially processed food out of their kitchens and bodies—than about fair wages and labor abuses. To the extent that the "good food" movement focuses on the choices that consumers make in the marketplace, it risks diverting attention from structural issues like the conditions of production in the food industry. It may also come across as sanctimonious and elitist. In many poor urban areas, farmers markets and grocery stores that sell organic chicken and grass-fed beef do not exist, after all, and their products would be unaffordable to many residents even if they did. The price of the organic meat sold at supermarkets like Whole Foods is significantly higher than the price of the value pack sold at Walmart. The bill at a typical farm-to-table restaurant—restaurants modeled on the locavore idyll, where everything from the heirloom vegetables to the grass-fed lamb is locally sourced—is likely to be double or triple the cost of a meal at a local diner, to say nothing of a fast-food restaurant.

It is easy enough to engage in virtuous consumption if you have plenty of disposable income. It is significantly harder for families living on tight budgets, much less those who rely on food stamps. The result is to create a virtue divide that all too often mirrors the class divide. While the poor buy the bad meat sold at KFC and Walmart, the rich consume the ethical meat sold at fancy restaurants and places like Whole Foods, purchasing beef and chicken adorned with labels that affirm their own sense of purity and virtue. As in so many other areas of life, virtue correlates with privilege, enabling affluent consumers to buy their way out of feeling complicit in the impure, dirty practices that go on inside factory farms and industrial slaughterhouses, which churn out food that other, less virtuous people consume. Among the less virtuous people are the workers who toil in these slaughterhouses, the live hangers and belly rippers who keep the "disassembly lines" running.

PART IV

THE METABOLISM OF THE MODERN WORLD

8

Dirty Energy

On the morning of April 21, 2010, Sara Lattis Stone began frantically calling the burn units of various hospitals in Alabama and Louisiana. She was searching for news about her husband, Stephen, who worked on an offshore oil rig in the Gulf of Mexico where a massive explosion had occurred. The blast took place the day before Stephen was scheduled to return home from his latest three-week hitch on the rig, a semisubmersible floating unit called the Deepwater Horizon.

In the hours after a spokeswoman from Transocean, the company that owned the Deepwater Horizon, called to tell her that an "incident" had required the rig to be evacuated, Sara veered between panic and denial. One minute, she was telling herself that Stephen was fine. The next, she was convinced that she would never see him again. On Facebook, she came across frightening messages—"the water's on fire!", "the rig is burning"—posted by the spouses of other workers. At one point, Sara got on the phone with one of them, a woman who had her TV tuned to the same channel that she was watching, which was airing live coverage of the blowout. As they peered at the screen, both heard the same update, describing the blast as a catastrophic accident and raising the possibility that no one on the rig had survived. The news prompted both of them to drop their phones and scream.

Sara lived in Katy, Texas, a town just west of Houston where she'd grown up and where she and Stephen had settled after getting married. The day after he got home from his hitch, they were planning to meet with a real estate agent, having just gotten preapproved for a loan to buy a house. Now Sara wondered if Stephen would ever come home. None of the hospital burn units that she tried reaching had any information about him. Eventually, she received another call from Transocean, informing her that although the blowout had caused multiple fatalities, Stephen was among those who'd managed to escape from the burning rig. The survivors were now being transported by ferry to a hotel in New Orleans, she was told. After consulting with her mother, Sara tossed some belongings into a suitcase, drove to the Houston airport, and boarded the next available flight to the Gulf. The following morning, at around 3:30, she got a call from Stephen, who told her he was on his way to the hotel where she and other family members had gathered to wait. "Are you okay?" she asked him. "Yeah, I'm fine," he said.

Later, when she saw him shuffle through the hall that had been cordoned off for surviving crew members, she knew immediately that he wasn't fine. His expression was blank, and like the other survivors, he looked shell-shocked and traumatized.

"When he walked in, from the look in his eyes, it was obvious that something horrible had happened," she recalled.

GRIMY CARYATIDS

Eating meat produced in industrial slaughterhouses is one way that consumers benefit from dirty work performed in distant places on their behalf. Relying on fossil fuels that are drilled, mined, and fracked to sustain their lifestyles is another. In 1937, after visiting the coalfields of Yorkshire and Lancashire the year before, a British writer named Eric Arthur Blair, better known by his pen name, George Orwell, reflected on society's dependence on the people who extracted these resources from beneath the earth. What Orwell found after descending into the pits—"heat, noise, confusion, darkness, foul air, and, above all, unbearably cramped space"— struck him as a "picture of hell," teeming with miners whose exertions were as invisible as they were essential to society. "In the metabolism of the Western world the coal-miner is second in importance only to the man who ploughs the soil," Orwell wrote in *The Road to Wigan Pier*. "He is a sort of grimy caryatid upon whose shoulders nearly everything that is *not* grimy is supported."

"Practically everything we do, from eating an ice to crossing the Atlantic, and from baking a loaf to writing a novel, involves the use of coal, directly or indirectly," Orwell went on. When you see workers stoop down and shovel coal onto conveyer belts inside the narrow, dust-choked tunnels, "it is brought home to you, as you are watching, that it is only because miners sweat their guts out that superior persons can remain superior. You and I and the editor of the *Times Lit. Supp.*, and the Nancy poets and the Archbishop of Canterbury and Comrade X, author of *Marxism for Infants*—all of us *really* owe the comparative decency of our lives to poor drudges underground, blackened to the eyes, with their throats full of coal dust, driving their shovels forward with arms and belly muscles of steel."

In Orwell's day, the griminess of coal mining—the ash and dust, the foul air—was physical, staining the clothing, as well as the faces and bodies, of the workers who ventured underground. ("The most definitely distinctive thing about them is the blue scars on their noses," Orwell wrote. "Every miner has blue scars on his nose and forehead, and will carry them to his death.") By the time Stephen Stone found himself on the Deepwater Horizon, the taint of working in the extraction industry was less physical than moral. People who cared about the environment associated the oil industry with events like the 1989 *Exxon Valdez* oil spill, which blackened the shorelines of Prince William Sound, and with carbon emissions that imperiled the planet. It was an industry whose pipelines and projects threatened delicate ecosystems like the Arctic National Wildlife Refuge in Alaska; an industry from which more and more reputable institutions—universities, philanthropic organizations—had begun divesting; an industry that anyone concerned about the fate of the earth would sooner protest than turn to for employment. JOBS FOR CLEAN ENERGY, NOT FOR DIRTY OIL, read a sign at a rally on a college campus in Iowa, expressing a view that more and more environmentalists, scientists, and young Americans shared.

But while condemning the greed of oil companies was easy enough, avoiding relying on the product they produced was more difficult, the evidence suggested. For all the talk of shifting to wind and solar power, fossil fuels still supplied 84 percent of the world's energy in 2019, and in many places their use was increasing. Part of the reason for this was the emergence of a middle class in countries like China and India. Another factor was the massive carbon footprint of the United States, which made up less than 5 percent of the world's population but consumed a quarter of the world's energy. More than eighty years after *The Road to Wigan Pier* was published, "dirty oil" was no less important in the metabolism of global capitalism than coal had been in Orwell's time, thanks in no small part to the lifestyles of Americans and to the policies of their leaders. Although he spoke frequently about the importance of addressing climate change, Barack Obama presided over a massive increase in crude oil production, which grew by 3.6 million barrels a day during his tenure. When Obama left office, the United States was the world's leading petroleum producer.* His successor, Donald Trump, was an even more unabashed promoter of the fossil fuel industry, rolling back environmental regulations to restore America's "energy dominance."

Stephen Stone did not grow up dreaming of working in the energy industry. He was far more interested in enjoying his natural surroundings. Throughout his childhood, his favorite place to spend time was outdoors, swimming in the Tennessee River or trekking through the wilderness near his home in Grant, Alabama, a small town nestled in the foothills of the Appalachians. The bucolic setting suited him, at least until he got a bit older, when life in a backwoods town with limited opportunities began to feel stifling. During what would have been his senior year in high school, he started working the night shift at a rug factory in nearby Scottsboro, the same factory where his mother worked after his parents got divorced. After graduating, he quit the rug factory and enlisted in the navy. Two and a half years later, in 2007, he was discharged, mainly because he'd spent too much time drinking and partying at the string of sun-splashed naval bases (Aruba, Panama City) where he was stationed. Upon returning to Grant, he started calling various oil companies to see if he could land a job on a rig, both because he'd heard that oil companies liked to hire former navy guys and because the work paid well, far more than any other job a high school graduate from rural Alabama was likely to stumble

---

* The measure combined oil and natural gas; in oil alone, America ranked third, behind Russia and Saudi Arabia.

across. Some time later, he flew to Houston to interview for a position as a roustabout with GlobalSantaFe, an offshore drilling company that would later be bought by Transocean.

It was on this visit to Houston that Stephen decided to strike up a conversation with the redhead sitting next to him on the airport shuttle. The redhead was Sara, who had just come back from a trip to the Sundance Film Festival in Park City, Utah. They chatted for three hours, bonding over everything from their shared southern heritage—he was from Alabama, she from Texas—to their fathers' fondness for the same restaurant in Houston, a clam-and-oyster bar called Captain Tom's. Afterward, they exchanged phone numbers. Within a year, they'd gotten married.

In some ways, Sara and Stephen made for an odd couple: she was a college graduate with an introspective manner; he was a good old boy who was quick with a joke and liked to laugh and party. From the moment they started talking, though, Sara was struck by Stephen's intelligence, the books he mentioned reading, and the thoughtful gaze in his eyes. Whenever he would go offshore on a hitch in the years to come, Sara would notice, Stephen made sure to pack some reading—novels, poetry, philosophy. He also brought along a couple of pocket-size notebooks that he would fill with poems and drawings. To some college graduates, marrying a rig worker, even one who wrote poetry in his spare time, might have seemed like a step down. To Sara, it felt natural. Virtually everyone she knew in Katy came from a family with ties to the oil industry. Her own father had worked in the industry for decades, which was another thing she and Stephen had in common. The rhythm of the lifestyle, marked by two- and three-week hitches during which rig workers were separated from their spouses, was familiar to Sara, who often went months without seeing her father during her childhood. When Stephen would leave on hitches, she would miss him, but she also liked having time to focus on her own interests, in particular her art. In college, she'd majored in painting and photography, visual mediums through which she'd always found it easier to express herself than words. After she graduated, one of the first jobs she landed was as a fine art reproduction artist, copying paintings that were sold to doctors' offices and furniture stores like Ethan Allen. The pay was modest, and duplicating other artists' compositions felt strange to her, but the job boosted her confidence and made her realize how important it was for her to find an outlet for her creative impulses.

In the aftermath of the explosion on the Deepwater Horizon, Sara's creative impulses went into overdrive. After seeing images of the spill on television, she persuaded Stephen to take a road trip through the Gulf so that she could film what was happening and take some pictures. She also started painting more seriously. Among the canvases she composed was a series of portraits of the blast's survivors. The paintings were drafted, fittingly, in oil and were inspired by a visit that she and Stephen paid to Washington, D.C., a few weeks after the blowout, where they and other survivors were invited to testify at a House Judiciary Committee hearing on the Deepwater disaster—a disaster that was still unfolding and that, upon closer inspection, was hardly a surprise. The immediate cause of the blast on the Deepwater Horizon was a bubble of methane gas that floated up through the drill column, most likely because of a breach in the cement casing that enclosed it, and spread across the deck before igniting into a deadly fireball. In the view of many analysts, the deeper cause was a lack of attention to risk—and an excessive focus on profits—that characterized the entire oil industry and was particularly pronounced at BP, the company that leased the rig from Transocean and owned the exclusive rights to the Macondo Prospect well, an oil and gas reservoir located forty-nine miles off the coast of Louisiana. In the 1990s, BP had undergone a restructuring, outsourcing many technical functions and concentrating on maximizing production in each of its so-called strategic production units. The company's emphasis on the bottom line, affirmed in the motto "Make every dollar count," drew praise from business analysts. Safety experts were more alarmed. In 2005, an explosion at a BP refinery in Texas City killed fifteen workers and injured hundreds more. An investigation conducted afterward by the U.S. Chemical Safety Board found "serious deficiencies" in "safety culture" at the refinery and faulted BP for pushing for 25 percent budget cuts "even though much of the refinery's infrastructure and process equipment were in disrepair." Between 2007 and 2010, OSHA cited BP for 760 safety violations, by far the most of any major oil company. Leasing the Deepwater Horizon cost BP one million dollars a day, and the Macondo well had fallen behind schedule, ratcheting up the pressure to brush aside concerns that might have slowed the pace of drilling. Some workers feared that raising such concerns would get them fired, which helps explain why an array of ominous signs—problems with the cementing, flaws in the blowout preventer—were ignored. Hours before the rig went up in flames, a BP executive on the rig congratulated the crew for seven years without a "lost-time incident." After the blowout, BP scrambled to contain the oil gushing out of the well, which remained uncapped for eighty-seven days, blackening and befouling everything it touched. Tar balls washed up on the shores of beaches in multiple states. Shrimp and oyster harvesters were put out of business. By the time the Macondo well was shut, an estimated 210 million gallons of oil had leaked out, twenty times the volume of the *Exxon Valdez* oil spill.

There were also human costs, which Sara sought to capture in her paintings. One of them was of Chris Jones, whose brother, Gordon, was one of eleven workers killed in the disaster. Sara sat next to him during the congressional hearing. In Sara's portrait, Jones's lips are pursed and his face, painted ash blue, is creased with anguish. Another painting depicted a woman with her mouth agape and tears shimmering in her bright blue eyes. It was a portrait of Natalie Roshto, whose husband, Shane, also died on the rig. Titled *Survivors*, the paintings were stark and vivid, capturing the raw grief that filled the room. But the portrait that Sara drew of Stephen captured something different. Based on a photo that was taken during his testimony, it shows a bearded figure with a vacant, faraway expression in his eyes. He does not look grief stricken so much as bewildered and unmoored.

The bewilderment was still apparent when I met Stephen several years later, at a bar in San Clemente, California, where he and Sara were living at the time. They were staying in an apartment that Sara's parents had bought (her father grew up in California and was planning to move there after retiring). Stephen was in his late twenties, with a shaggy mop of chestnut-colored hair and languid, downcast eyes. At the bar, he was taciturn, nodding occasionally at something Sara said while straining to keep his gaze from drifting off. Unlike some of the workers on the Deepwater Horizon, he had managed to escape from the rig without sustaining any burns or physical injuries. But as I would come to learn, the absence of visible wounds was a mixed blessing, prompting friends to wonder what was wrong with him and exacerbating the shame he felt for struggling to move on. Since the explosion, he'd been unable to hold down a job. He avoided social gatherings. He also had trouble sleeping. Not coincidentally, the explosion on the rig had happened at night, collapsing the stairwell above the room in which Stephen had fallen asleep after completing a work shift. The blast startled him awake and sent him racing into the change room, where he slipped on a pair of fire-retardant coveralls and fumbled his way toward the deck, at which point he saw that the entire rig was smoldering and heard the panicked screams of his coworkers. It was an experience he now feared reliving every time he shut his eyes, Sara had come to realize. "The way I understand it is, he's constantly preparing for that wake-up," she said.

In the days that followed, I visited Stephen and Sara several times in their apartment, a two-story dwelling in a complex of look-alike gray bungalows where they lived with a terrier named Kale and two pet ferrets. With each successive visit, Stephen grew a bit more open and talkative. He told me about his upbringing in Grant. He described his stint in the navy. He recited some of his poetry and showed me his collection of books—Shakespeare, Yeats, Thoreau. The blowout had not diminished his appetite for reading, but the genre that captivated him had changed. He now gravitated mainly to science fiction and had gotten particularly obsessed with outer space, which Sara interpreted as a sign of the compulsion he felt "to be away from reality, away from earth." Stephen did not dispute this interpretation. He also did not deny that after the blowout another way he'd tried to escape from reality was by consuming large amounts of alcohol. The liquor helped him fall asleep at night, but it also fueled some erratic behavior. One night, after drinking with a friend, he grabbed the keys to their car, zoomed down a one-way street in the wrong direction, and rammed into a brick wall. The accident fractured vertebrae in his neck and collapsed a lung. When we met, Stephen had cut back on the drinking, but an air of melancholy enveloped him. Much of the time, so did a plume of marijuana smoke. Throughout my visit, he sat on an L-shaped couch in the living room of the apartment, sipping black coffee from a green mug and, every few minutes, lighting up and taking another toke from a bowl of weed. The pot was medical marijuana that a psychiatrist had prescribed to quell his anxiety. The same psychiatrist had diagnosed him with PTSD.

Given what he'd been through—a near-death experience that shattered his sense of security—this diagnosis made sense. Like military veterans who'd survived IED explosions in Iraq, Stephen was sensitive to loud noises and given to paranoid fears and panic attacks. The rattle of ice in the freezer was enough to set him off sometimes, Sara told me. A few days before I arrived, she'd found a knife on the dashboard of their car. Stephen had pulled it out after becoming convinced that the driver he'd spotted in the rearview mirror was following him.

But as with many military veterans, there was something else that seemed to afflict Stephen no less: not fear but anger and disillusionment. These feelings percolated immediately after the blowout, he told me, when the rig's survivors arrived at the hotel in New Orleans. They were exhausted and still reeling from the shock, yet before getting to see their families, Stephen said, they were taken to a meeting room where a Transocean manager delivered a speech that sounded to him like an exercise in spin control. The experience left a bad taste in Stephen's mouth. A few weeks later, a Transocean representative reached out to him and, over a cup of coffee at Denny's, offered him five thousand dollars for the personal belongings he'd lost on the rig, which he accepted. Then the representative asked him to sign a document affirming that he had not been injured. Stephen was dumbfounded. "I'm not signing this," he told the representative. "I don't know if I'm injured yet—this just happened."

To both Transocean and BP, the survivors were not human beings who deserved to be treated with dignity, Stephen was coming to feel, but a potential legal liability that needed to be contained. To people with a cynical view of the oil industry, this would not have come as a surprise. But Stephen was not such a person. When he applied for the job at Transocean, he understood that working on an offshore oil rig could be dangerous, but he also assumed the industry did everything possible to protect workers. "I thought everything was followed to a T safety-wise," he said. After the blowout, as he read about how little money oil companies spent on safety and how many warning signs on the Deepwater Horizon had been ignored, a wave of disillusionment washed over him. "Who the fuck was I working for?" he wondered.

Stephen wondered this again when, within a year of the Deepwater Horizon disaster, Transocean awarded lavish bonuses to several senior executives for overseeing the "Best Year in Safety Performance" in the company's history. The decision was announced only a few months after a nonpartisan national commission submitted a report about the Deepwater spill to President Obama. Based on an exhaustive six-month investigation, the report linked the blowout to "a series of identifiable mistakes made by BP, Halliburton, and Transocean that reveal such systemic failures in risk management that they place in doubt the safety culture of the entire industry." The report listed nine questionable decisions that were made when less risky alternatives were available, alternatives rejected because pursuing them might have cost money and time. When Stephen learned about Transocean's "Best Year in Safety Performance" bonuses, he was still a Transocean employee. Afterward, he submitted an angry resignation letter. "I quit," he said. "I was like, fuck you guys. I don't want to be a part of your company."

In *Achilles in Vietnam*, the psychiatrist Jonathan Shay argued that a primary cause of moral injury among military veterans was the betrayal of "what's right" by their commanders, giving rise to *mênis*, a Greek term that Shay likened to "indignant rage . . . the kind of rage arising from social betrayal that impairs a person's dignity." According to Shay, this is what consumes Achilles after his commander, Agamemnon, violates his sense of moral order in the *Iliad*. It is what Shay found again and again among veterans who felt their dignity had been trampled on in Vietnam. And it is what appeared to grip Stephen, who felt deeply betrayed by an industry that upended not only his sense of security but also his moral bearings and his trust.

"I think there's the personal betrayal of the company-employee relationship," he said. "But there's an even larger sense of betrayal. I didn't think the industry was this bad." He paused. "It just kind of takes some hope from humanity, shatters your illusions a little bit."

There was one other betrayal that appeared to weigh on Stephen: the betrayal of himself, the part of him that loved nature and, after the blowout, as the scale of the disaster became clear, felt dirtied and implicated.

He felt this in particular on the road trip that Sara persuaded him to take through some of the places in the Gulf where the pollution from the spill had begun to wash up. Among their destinations was Dauphin Island, on Alabama's Gulf Coast. During his childhood, Stephen had vacationed there with his family. It was one of his favorite places, famous for the ribbon of pristine white sand that graced its shores. After the Deepwater spill, much of the island was surrounded by orange booms, and the sand was stained with oil sludge, a sight that filled Stephen with both shame and sadness. "This great place from my childhood was getting shit on," he said, "and I was part of the group that shit on it."

For Sara, too, seeing the impact of the spill dredged up difficult feelings about the world she'd grown up in, a world she realized she had long viewed through rose-tinted glasses. Throughout her upbringing, she told me, she was taught that oil rigs were actually good for the environment because, when they sank, they created reefs for fish. The footage she recorded on the trip through the Gulf told a different story. Later, when she watched BP air ads on television burnishing its commitment to the environment, she was furious. But unlike Stephen, Sara bristled at blanket condemnations of the oil industry that failed to distinguish between companies that behaved recklessly and those that at least tried to act responsibly. She also bristled at environmental groups that, after the blowout, seemed to focus far more attention on the pelicans and dolphins poisoned by the oil spill than the rig workers who'd died. Every day on the news, it seemed, she would see images of dead seabirds and marine mammals. The faces of the rig workers never appeared. Sara could not understand why they were so invisible. "It's just weird," she said.

But Stephen did not seem to find it so weird. Most of the people he worked with were "blue-collar guys" and "country bumpkins" from backwoods towns like the one he'd grown up in, he noted. The kinds of people "superior persons" looked down on, in other words. Then he mentioned another reason why the public might find it easier to sympathize with dead dolphins than with workers like him.

"People see the environment as completely innocent," Stephen said, "whereas we, just being in that industry, you know, you kind of brought it on yourself."

"TROUBLED WATERS"

Stephen did not seem to begrudge people for feeling this way, and arguably with good reason. He had, after all, collected a paycheck from Transocean, making upwards of sixty thousand dollars a year as a roustabout, a salary that was bound to increase as he gained more experience (roustabouts were entry-level deckhands who were paid less than anyone on rigs save for the mess hall chefs). Were it not for the blowout, he probably would have continued working in the industry, he told me, for the same reason most of the blue-collar guys on the Deepwater Horizon did: the money was good. The same incentive explained why thousands of working-class men flocked to places like the Williston Basin, home to the Bakken Formation, during the fracking boom, where drillers and riggers could sometimes pocket more than ten thousand dollars a month. Some of Stephen's coworkers on the Deepwater Horizon earned six-figure salaries despite having nothing more than a high school diploma. As with fracking, the job was hard—twelve-hour shifts during which Stephen raced around stacking equipment, mixing drilling mud, and performing other menial tasks. But it beat living paycheck to paycheck with few benefits or vacations like everyone he knew back in Grant. When Stephen came home from hitches, he and Sara would often hit the road, taking trips to places like Zion National Park in Utah without worrying about when the next paycheck would arrive.

"A path to a life otherwise out of reach" was the phrase that a team of reporters from *The New York Times* used to describe how the crew members on the Deepwater Horizon viewed their jobs, a life with its share of perks and benefits. If environmentalists had little sympathy for the workers who enjoyed these perks while ignoring the "dirty facts" about the fossil fuel industry cataloged in a report published by the Natural Resources Defense Council—water pollution, land degradation, the discharge of three-fourths of America's carbon emissions—who, really, could blame them? These dirty facts were real, Stephen acknowledged. On the other hand, it was not lost on either him or Sara that a lot of people who saw rig workers as complicit in these dirty facts were happy enough to pump gasoline into their SUVs and minivans without feeling the least bit sullied themselves. "We like to forget that our everyday lives are what's making that the reality," Stephen said of an industry that catered to America's insatiable demand for cheap oil, just as companies like Sanderson Farms and Tyson catered to the demand for cheap meat.

Who ended up doing this work was shaped by class but also, perhaps even more so, by region and geography. In 1994, the sociologists William Freudenburg and Robert Gramling examined the regional patterns in a book titled *Oil in Troubled Waters*, which compared the status and prevalence of offshore drilling in two states with large shorelines, Louisiana and California. It was in California that in 1969 a blowout on an oil platform in the Santa Barbara Channel first drew attention to the environmental risks of offshore drilling. The spill sparked widespread outrage, galvanizing support for the National Environmental Policy Act, a landmark law that was enacted the following year. It also prompted the then secretary of the interior, Walter Hickel, to issue a moratorium on offshore drilling in California's waters. Decades later, few residents of the Golden State were clamoring to change this, Freudenburg and Gramling found. Virtually every Californian they interviewed opposed offshore drilling. The opposition was so uniform that they started to ask subjects if they simply *knew* anyone in California who held a different view. Virtually no one did.

In southern Louisiana, a series of blowouts also took place in the early 1970s, polluting the Gulf and, in some cases, causing fatalities. But unlike in California, no moratorium on offshore drilling ensued. By the time Freudenburg and Gramling conducted their study, more than thirteen thousand production wells had been drilled in the Gulf of Mexico's outer continental shelf. Once again, the subjects of their study all seemed to hold the same view of this activity, only this time it was the *opposite* view: in Louisiana, opposition to offshore drilling was nonexistent. Only a handful of respondents knew someone who harbored even mild reservations about it.

What accounted for these starkly divergent attitudes? One explanation was ideological: California was a liberal state whose residents tended to care about the environment and distrust industry, whereas Louisiana was a conservative one where people held favorable views of business.

But the divergence also reflected radically different economic prospects. As Freudenburg and Gramling noted, the Californians they interviewed did not seem to care that closing the coast to drilling might hamper economic development. In fact, many of them were transplants from other states who had chosen to live in Northern California "to get *away* from that kind of shit," as one respondent put it, describing rigs and derricks as eyesores that would defile the state's natural beauty, which needed to be protected from development. Louisianans did not have the luxury of thinking this way. The oil industry meant jobs in a place that was starved of them. To no small extent, this belief helped *explain* the ideological differences, Freudenburg and Gramling's study suggested.

More than a decade later, the Berkeley sociologist Arlie Russell Hochschild discovered something similar when she began interviewing Tea Party activists in Louisiana's bayou. Like her neighbors in Berkeley, Hochschild's subjects did not like the idea of eating oil-tainted shrimp and seeing their lakes and rivers polluted. But many saw drilling as essential to their survival. "The more oil, the more jobs," went the logic Hochschild encountered. "The more jobs, the more prosperity." Which, in turn, led people to support luring oil companies to Louisiana by offering them lower taxes and less government regulation. Unlike Freudenburg and Gramling, Hochschild did not find that support for drilling was monolithic. Younger, college-educated Louisianans who lived in urban areas often held different views. But among longtime residents of southern Louisiana who lived in smaller towns and had less education, it was extremely pervasive. As it happens, these residents fit the demographic described in a consulting report Hochschild came across, advising companies that owned plants or refineries that emitted large amounts of pollution on where to locate. According to the report, the ideal location was a place with a high concentration of "least resistant personalities." One characteristic of "least resistant personalities" was that they were longtime residents of small towns. Another was that they had only high school educations.

It was easy for the residents of liberal, eco-conscious states to view such people with condescension while conveniently forgetting how dependent on them they were. "The centrality of oil and gas exploration to the Gulf economy is not widely appreciated by many Americans, who enjoy the benefits of the energy essential to their transportation, but bear none of the direct risks," noted the national commission report on the Deepwater Horizon disaster. People in California could essentially off-load these risks to less privileged Louisianans who did the dirty work of running petro-chemical plants and drilling offshore for them. And yet the truth is that even in southern Louisiana energy companies were not always welcomed with open arms. Back in the 1930s, fishers and trappers in the Bayou region expressed bitter resentment at the *maudits Texiens* (damn Texans) who began dredging canals and drilling holes into salt domes to extract oil from beneath the land that was their main source of livelihood. "It was like we had been invaded," one resident complained. Over time, however, the ill will gave way to pragmatism. The opportunity to work in the oil in-dustry came to be appreciated in places like Black Bayou, home to an oil field run by Shell, and later, as companies shifted their focus to offshore drilling, in the towns and parishes dotting Louisiana's coast.

By the end of the 1990s, nearly one-third of America's domestic energy supply came from offshore production in the Gulf of Mexico. To Louisi-anans who found jobs in the petroleum industry, this was both a source of livelihood and a point of pride, even if serving as "America's Energy Coast" also had downsides, including the highest level of air pollution in the country and the degradation of Louisiana's coastal wetlands. In *American Energy, Imperiled Coast*, Jason Theriot, a historian who grew up in southern Louisiana and was the grandson of an oil worker, described how, in the late 1990s, some Louisiana politicians began to argue that their state deserved more help to offset these costs. Much of the oil and gas that flowed through the state's pipelines ended up servicing other parts of the country, they pointed out, absorbed into the metabolism of prosperous regions like New England. Meanwhile, Louisiana's coastal communities were sinking, leaving the residents of cities like New Orleans more vul-nerable to storms and hurricanes, a problem likely to grow worse in the future, thanks to rising sea levels precipitated by climate change. Loui-siana "has not received appropriate compensation for the use of its land and the environmental impacts of this production," argued Senator Mary Landrieu, who introduced a bill, the Conservation and Reinvestment Act, that called for her state to receive more of the revenue generated by offshore drilling to help restore its wetlands. Notably, however, the leaders pushing for reinvestment stopped short of threatening legal action, much less of calling for a moratorium on offshore drilling, steps that risked alienating an industry on which Louisiana relied. As Theriot observed,

"Those leaders carefully walked the fine line between pursuing the restoration agenda and preserving the main driver of the state's economy."

The desire to preserve this driver was understandable in a state that ranked among the poorest in the country. As a way to foster long-term prosperity, it was also likely to be futile. While boosters hailed energy production as a catalyst for growth, studies showed that communities that relied too heavily on resource extraction often suffered from persistent poverty. Such communities were vulnerable to dislocations when prices fell. Many also underinvested in human capital, a problem exacerbated by industry's demand for tax breaks, which made it hard to fund schools and other public institutions. A striking example of this was Cancer Alley, an industrial zone in Louisiana that was home to 150 refineries and chemical plants, facilities that belched out vast quantities of pollution but that often avoided paying taxes, which was why the quality of the public schools was no better than the air quality. "Without property taxes, these parishes do not have money for their schools," Cynthia Sarthou, the executive director of the Gulf Restoration Network, told me. "They get little or no money and *all* the pollution." In the rest of Louisiana, it wasn't a whole lot better, said Sarthou, who likened the state's dependence on oil to an abusive relationship in which the victim kept coming back because she feared it was all she had. "We get the promise of income, so we put up with it," she said. "Yet we're still one of the poorest states in the nation."

A few days after meeting Sarthou, I saw some of this poverty up close on a visit to Morgan City, a port town where, in 1947, an offshore well was drilled beyond the sight of land for the first time. A plaque commemorating this milestone had been erected on the median of the first road I turned onto, next to a sculpture of an oil rig that looked weather-beaten and oxidized, with patches of rust spreading over its corroded metal facade. In the 1970s, after the OPEC embargo caused the price of oil to rise, Morgan City had been a boomtown, its harbor jammed with rigs and its streets crowded with people who came looking for jobs. So many outsiders flocked to town that "roustabout camps" sprang up to accommodate them. But by the mid-1980s, oil prices had fallen and the boom petered out. When I visited, the roustabout camps were long gone, and the only rig I saw in Morgan City's harbor was Mr. Charlie, a rusted yellow pile that was now the main attraction at the Rig Museum. It was located at the end of a gravel road abutting the harbor and offered tours to visitors curious to stroll aboard an authentic oil rig. On my tour, which a couple of French tourists joined, a guide explained that Mr. Charlie had stopped drilling wells in 1986 and would have been converted into scrap metal if it hadn't been saved by a local man named Virgil who decided to turn it into a tourist attraction. A modest attraction, to judge by the size of our entourage and the humble quarters of the museum, which was housed in a trailer park. After the tour, I chatted with Bryce Merrill, a former oil industry worker who was now the museum's curator. Merrill didn't hide the fact that Morgan City had seen better days. "There used to be seventy rigs out here," he said, pointing to the harbor. "Now there are five. We've lost twenty-five hundred people in the last four years, and about twenty-five to thirty businesses have closed."

After stopping for lunch, I paid a visit to James Hotard, a short, barrel-chested man who worked as a training manager at Oceaneering, which supplied subsea hardware to offshore rigs. Hotard recalled the time when virtually anyone in Morgan City could find a job. "You didn't even have to graduate high school," he said. "Just get to eleventh grade, say I'm done with school, step out, and you could make forty to fifty thousand dollars as a welder." Nowadays, he told me, young people with ambitions went to college and, upon getting their diplomas, moved elsewhere (his own daughter had just graduated from LSU and was heading to Houston). Driving through Morgan City, I didn't find it hard to fathom why. Empty storefronts lined the moribund commercial center, which looked less like a boomtown than a ghost town. The housing on the residential streets was in a state of dramatic disrepair—porches crumbling, roofs moldering, wooden beams jutting out of run-down shacks. In front of one of the houses, I spotted a weather-beaten basketball hoop with a shredded net hanging from the rim. Next to it was an empty shopping cart. The image reminded me of something I'd seen before, though I couldn't recall exactly what. Later, it came to me—the visit I'd paid to Florida City to see the Dade Correctional Institution, during which I'd stopped to talk to Jimmy, the man who was selling mangoes and lychees on the street, next to an empty shopping cart.

"THE SHARP END OF THE SPEAR"

Just as dirty work was disproportionately allotted to certain classes, so, too, was it concentrated in certain places: "rural ghettos" where prisons like Dade were located; remote industrial parks where slaughterhouses set up shop; towns full of "least resistant personalities" where refineries were built and coastlines were opened to offshore drilling, creating eyesores that the residents of places like Marin County, California, were spared. The geography of dirty work both mirrored and reinforced race and class inequality, ensuring that stigmatized industries and stigmatized institutions were situated in less affluent parts of the country, places like "Cancer Alley."

To be sure, not everyone saw these industries as stigmatized or regretted working in them. A few days before visiting Morgan City, I had dinner with Rick Farmer, a drilling engineer. We met in La Villa Circle, a gated community near Lafayette where he lived in a large brick house with a columned porch and a pair of magnolias out front. Farmer had grown up in more hardscrabble circumstances. The son of farmers, he cut his teeth in the oil industry when he was eighteen, working as a roustabout while attending Mississippi State University, where he earned a degree in petroleum engineering. He had just turned sixty and was financially secure and openly grateful for a career that had enabled him to avoid the drudgery of a conventional day job. Working offshore was "an adventure," he said. He loved the feeling of being out at sea and the rush that came from getting things done. He also didn't mind the salary, mentioning that he made between $350,000 and $400,000 a year.

If this was dirty work, a lot of Americans would have happily signed up for it, I thought while listening to Farmer describe how lucky he felt in the brick-floored kitchen of his spacious home. But not everyone would have considered him lucky, nor had he always felt this way himself. In 1984, he was pulling a well apart when some equipment collapsed. The falling debris instantly killed one of his coworkers and left Farmer permanently disabled, bound to the wheelchair in which he sat during dinner. His house was designed to accommodate his disability, with halls and doorways that were wide enough for him to maneuver around. This had not always been the case on the rigs he'd worked on. At dinner, he recalled having to crawl on his hands and knees to go to the bathroom. He also re-called the despondency he felt after the accident, when he took to drink-ing to try to dull the pain. When the pain persisted, he contemplated suicide. "Two years after it happened, I was ready to end it," he said. Eventually, some friends persuaded him to go to a rehabilitation center, where he met with a psychiatrist who handed him a pillow that he tried with all his might to tear apart. "Why me?" he howled. Then Farmer, who'd been raised Catholic, flung aside the pillow and thought of the torment endured by Christ during the Crucifixion. "We are all forsaken," he told me he realized in the throes of his despair. He had been a devout churchgoer ever since.

Surviving a near-death experience was not unusual in the oil industry. The person who'd introduced me to Farmer was Lillian Espinoza-Gala. She lived in Lafayette, in a wood-frame house crammed with books about the oil industry, for which she began working in 1973, becoming one of the first female roustabouts in the Gulf of Mexico. At the time, there were no separate bathrooms for women on rigs, she said. There were also few jobs for which Espinoza-Gala seemed less suited. Although her fa-ther worked in the oil industry, she had drifted in a different direction, joining the peace movement after several of her high school classmates returned from Vietnam in body bags. At one antiwar rally, she handed out a pamphlet that read, "Big Oil Is Raping the Earth." Eventually, af-ter a series of adventures that took her all the way to Canada, where her relationship with a charismatic peace activist went awry, she returned to Louisiana and began looking for a job so that she could pay the tuition to go back to school. One day, her father came home and told her that offshore companies were going to start hiring women. "I could do that," she said. Some time later, she went on her first hitch, pulling her sandy-blond hair beneath a hard hat and trading in the jeans swathed in antiwar patches that she often wore for a pair of coveralls. All her friends in the peace movement were appalled. To her surprise, she actually liked the job. Working on an offshore rig was like "living on a castle in the middle of the ocean," she told me. There was a lot about the job she came to love.

What she did not love was seeing more people die for no good rea-son. On two occasions, she was on a rig when fatal accidents occurred. One of the accidents, in 1981, caused by corroded wellheads about which she had tried to sound the alarm, killed a welder and injured her severely. Afterward, she was helicoptered to the ER, blood spattered and unconscious. The accident shattered several bones in her right hand, which was permanently disfigured, and ended her career as a roustabout. But it eventually propelled her to embark on a new career, as an industry safety consultant determined to prevent other rig workers from going through similar ordeals.

To some extent, accidents on offshore rigs were unavoidable. But the toll in lives was not the same in all countries. Between 2004 and 2009, fatalities in the offshore industry were "more than four times higher per person-hours worked in U.S. waters than in European waters, even though many of the same companies work in both venues," noted the national commission report submitted to President Obama after the Deepwater Horizon disaster. The report traced this disparity back to the 1980s, when a series of deadly accidents took place, including a blowout on the Piper Alpha, a platform in the North Sea, that killed 167 people. In Norway and the United Kingdom, the response was to enact stronger regulations that put the burden of preventing future disasters on industry. The United States adopted a laxer approach. One reason for this was that the oil industry vigorously opposed regulation. Another was that the U.S. Minerals Management Service, the federal agency responsible for overseeing safety and environmental standards in the Gulf, also oversaw the collection of royalties from oil and gas leasing operations. After taxes, these royalties were the second-largest source of government revenue, which helped explain why the MMS approached industry "more as a partner than a policeman," as one agency official quoted in the national commission report put it. None of this changed as oil companies started drilling riskier wells in deeper waters. Nor did it alter much as different administrations took office. As the national commission's report noted, "The safety risks had dramatically increased with the shift to the Gulf's deepwaters, but Presidents, members of Congress, and agency leadership had become preoccupied for decades with the enormous revenues generated by such drilling rather than focused on ensuring its safety. With the benefit of hindsight, the only question had become not whether an accident would happen, but when."

Espinoza-Gala served as a consultant on the national commission report, helping to write the chapter on worker safety. The Deepwater Horizon disaster did lead to some positive changes, she told me. The MMS was replaced by the newly formed Bureau of Safety and Environmental Enforcement. New safety regulations were passed. Yet Espinoza-Gala was understandably in no mood to harp on this progress on the day we got together. We met on May 2, 2019, which happened to be the thirty-eighth anniversary of the accident that almost killed her. The day before, the Trump administration put out a revised well-control rule that gutted the safety regulations passed after the Deepwater blowout. Under the new rules, independent auditors were no longer required to inspect safety equipment on rigs. Testing requirements for blowout preventers were also weakened. Overseeing these changes was Scott Angelle, a close ally of industry whom Trump had appointed to run BSEE and who proceeded to meet repeatedly with executives from oil companies that had been cited for violating safety regulations. "Help is on the way," Angelle promised the executives. Espinoza-Gala did not mince words about what delivering on this promise would mean for rig workers. "He has the potential to kill people that are connected to this community, and it will be on his soul the rest of his life," she said of Angelle, who was from Lafayette. Yet she acknowledged that it might be wishful thinking to assume this would weigh on his conscience. Espinoza-Gala told me that years earlier, when Angelle was running for Congress, Espinoza-Gala told me that she started calling various candidates in the race to ask them what they were going to do about offshore safety. "Offshore safety?" one of the candidates responded. "That's not an issue, Lillian. There's no jobs! The only thing people care about in Louisiana is jobs."

Just as Louisianans did not have the luxury to protect their coastline from environmental degradation, so, too, did they lack the luxury to protect rig workers from getting maimed and killed, the candidate was essentially telling her. Espinoza-Gala refused to accept this logic, both because of her own near-death experience and to honor the memory of the eleven workers who died on the Deepwater Horizon. When the blowout happened, she told me, it felt just like 9/11 to her. Ironically enough, it also led her to conclude that the emphasis on safety was misplaced—*personal* safety, that is. "Personal safety is, Wear your hard hat! Don't fall down!" she explained. It was the message that companies drummed home to workers in training sessions, leading them to think that avoiding accidents was up to them, which is what she herself had believed as a roustabout. "When I worked offshore, I thought all our accidents were caused by us," she said. It was the same message dirty workers in prisons and slaughterhouses received: if a problem existed, it was their personal fault. What mattered far more, Espinoza-Gala had come to believe, was "process safety," a product of choices made long beforehand—cutting costs, rushing projects—that made the entire system unsafe. Process safety flowed from decisions reached by senior executives who sat at what she called "the blunt end of the spear," as opposed to their underlings on "the sharp end of the spear," the frontline workers who risked death and injury, doing the dirty work for them.

The drillers and roustabouts imperiled did not all work offshore. According to Michael Patrick F. Smith, who went to North Dakota to get a job as a swamper during the fracking boom, "From 2008 to 2017, roughly the same number of oil field workers were killed on the job as U.S. troops in Afghanistan." Since the Deepwater disaster, Espinoza-Gala had been giving PowerPoint presentations about how to prevent more workers from dying, in part to debunk a common narrative that emerged after blowouts, which is that lives might have been saved if the people on the rig—the ones at the sharp end of the spear—had acted more responsibly. This narrative had important moral consequences, leading workers to blame themselves or their peers, rather than the executives pulling the strings above them, when things went awry. One afternoon when I visited her, Espinoza-Gala led me into her office, a small, wood-paneled room whose walls were decorated with various mementos. On one wall was an award recognizing her as "one of the first Gulf of Mexico female production roustabouts." On another was a picture of eleven wooden crosses planted on a strip of sand, one for each of the workers killed on the Deepwater Horizon. Espinoza-Gala slid into a chair in front of a computer monitor and, using the two fingers on her right hand that she could still bend freely, began clicking on a mouse pad. Halfway through her PowerPoint presentation, she clicked on a slide that showed the faces of each of the workers who had died. There was Donald Clark, forty-nine, an assistant driller from Louisiana. There was Aaron Dale Burkeen, thirty-seven, a crane operator from Mississippi. Wearing hard hats and safety gloves—personal safety—didn't help these workers, she noted. Feeling less pressure to keep costs down, or more freedom to report safety concerns to their superiors—process safety—might have. Some of the workers on the Deepwater Horizon had shared their concerns about the lack of safety with family members, Espinoza-Gala told me. One of the victims even asked his wife to make out a will for him just before the blowout. Other workers confided in Lloyd's Register, which surveyed conditions on the Deepwater Horizon a month before the disaster, interviewing forty workers, several of whom said they "often saw" unsafe practices but were too afraid to report them. After the blowout, Congress recommended extending whistleblower protections to rig workers to alleviate this fear. But in 2017, BSEE determined that enforcing these protections was beyond its purview, another gift to industry from Scott Angelle and the Trump administration.

Before shutting down her computer, Espinoza-Gala clicked on one other slide featuring a worker, a bearded man in a navy suit and silk tie who was sitting at a congressional hearing, delivering testimony. It was Stephen Stone. Behind him was a woman with long red hair and freckled cheeks dabbing a tear from her eye. It was Sara. On the next slide, a congressman was shown holding up a photo of one of the blowout's victims: a pelican, Louisiana's state bird, encrusted with crude oil. A proud Louisianan and committed conservationist ("Bayou Vermilion Watershed: Keep It Clean," read a sticker on the refrigerator in her kitchen), Espinoza-Gala was not unmoved by the image of the pelican. But, like Sara Lattis Stone, she found it difficult to understand why the pelicans aroused more sympathy from politicians than the workers. "The widows were in these hearings, where they're holding up pictures of birds instead of their husbands!" she said. For a long time, she told me, this enraged her. Eventually, she came to terms with it, reluctantly concluding that if not for the pelicans, the Deepwater Horizon disaster would likely have gone unnoticed in Washington, the way most rig accidents did, owing to the low value placed on the lives of the roustabouts who did the dirty work.

"If eleven workers would have died, nobody would have cared," she said. "It's only because of the birds and the pollution."

"LAYERED EFFECTS"

A few weeks after meeting Lillian Espinoza-Gala, I saw Sara Lattis Stone again. She was still living in California, but she was no longer married to Stephen. Not long after I'd visited them in San Clemente, they had moved to Portland together, seeking a fresh start in a calmer, slower-paced environment. The move took place shortly after they'd reached a settlement with Transocean that compensated them for the pain and suffering the blowout had caused. Both BP and Transocean ended up paying out billions of dollars in such settlements, leading some media outlets to portray the companies as victims of opportunistic lawyers filing a blizzard of inflated claims. Sara did not share the amount of their settlement with me. What she did tell me was that in the nearly five years it took to iron out, a period when Stephen couldn't work and she had to stay home to take care of him, they burned through all their savings and had to rely on family (in particular, her parents) to get by. She also told me that they wanted to hold out for a trial but ended up agreeing to a settlement out of desperation as medical bills went unpaid and debts piled up. "I feel really crappy about that," she said. For her and Stephen, the settlement was both a relief and a source of discomfort. Although the cash was welcome, they came to see it the same way Flor Martinez viewed the bonuses her husband earned as a supervisor at Sanderson Farms: as dirty money.

In Portland, their lives would return to normal, Sara had assumed. She and Stephen moved into a large craftsman-style house on a tree-lined street in a gentrifying neighborhood. Sara turned a room on the ground floor into an art studio. "Everything would be okay," she told herself hopefully. But things were not okay. Instead of calm, the neighborhood was noisy, with police sirens that blared through the night, setting Stephen's nerves on edge. Meanwhile, Sara began to experience a physical breakdown, migraines and ripples of pain that flared in her back and spread through her body. The pain grew so severe that, eventually, she couldn't get out of bed. "At one point, my legs were so heavy I couldn't walk up the stairs," she recalled.

Until this point, Sara had reserved most of her energy for caring for Stephen—making sure he took his medications, monitoring his intake of food and alcohol, assuaging his anxiety. Now her attention shifted to her own needs. To alleviate the pain in her back, she started seeing a chiropractor. To ease the emotional distress that she sensed might be causing it, she started seeing a therapist. The experience was transformative, enabling her to begin working through what she came to see as her "secondary trauma," a condition that she learned could afflict people who spent prolonged amounts of time in the company of individuals with PTSD and vicariously absorbed some of their symptoms. The condition was prevalent among the wives of military veterans, to whom Sara began reaching out on online chat forums. But as Sara took steps to take better care of herself, she also grew more aware of how unhealthy Stephen's coping mechanisms were. In San Clemente, he had spent his days reading sci-fi novels and smoking marijuana. In Portland, he discovered a new diversion, long, intricate board games that sometimes took weeks to solve, an activity that occupied his mind but pulled him further and further away from reality. The zeal he used to display for hitting the road and exploring new places after returning from hitches had not resurfaced. Neither had the levity Sara remembered from the pre-blowout days, when he would walk into a room and say something that instantly got people laughing and smiling. These qualities, she was beginning to suspect, might never resurface.

As this realization sank in, Sara decided to take a trip to Houston, a five-day vacation that was both a chance to see some friends and an opportunity to find out how Stephen would fare in her absence. On her third day away, she received a phone call, informing her that he'd gotten into another car accident after drinking too much. Although no one was hurt, the accident shook her profoundly, causing her to wonder what might happen the next time she left his side. The thought of living with this fear for as long as they were married frightened Sara, who had always valued her independence. A few weeks later, she told Stephen she wanted to split up.

Sara relayed all of this to me over dinner at a restaurant in Los Angeles, around the corner from the apartment in Hollywood to which she'd moved after she and Stephen separated. She had come to L.A. to make art—more specifically, to make movies. She was now a graduate student, enrolled in UCLA's acclaimed film school. The inspiration to apply to the program had come from one of the few new friends she'd made after the Deepwater blowout, a documentary filmmaker named Margaret Brown who had reached out to her back in 2011, after coming across her *Survivors* paintings. Impressed by the work, Brown asked Sara whether she and Stephen might be willing to appear in a film that she was making about the Deepwater spill. Under normal circumstances, the answer would have been no: after the blowout, neither she nor Stephen felt any desire to expose their lives to an outsider, much less an outsider pointing a camera at them. But Brown, who had grown up in Alabama, was a fellow southerner and fellow artist with whom Sara felt a connection. Her film, titled *The Great Invisible*, came out in 2014, deftly interweaving stories of the human beings whose lives were devastated by the spill—rig workers, oyster shuckers—with vignettes illuminating how quickly business as usual resumed once the disaster faded from the headlines. In one scene, a former chief mechanic on the Deepwater Horizon who was badly injured by the blowout talks about the less visible emotional wounds that he bears. "It makes me feel guilty 'cause I played along," he says as his wife prepares a slew of pain and anxiety pills for him to take. "A lot of things that I was doing, I knew were wrong. I feel really guilty for working for BP." In another scene, a group of executives attending an oil and gas trade show in Houston toast the industry's resurgence on the deck of a luxury hotel. After lighting up cigars, one of the executives says, "Personally, I think we should tax the living hell out of gasoline." "That's a political statement," another objects. The argument eventually goes the way of the smoke wafting over the table and the carbon emissions the oil industry belched out year after year, drifting into the atmosphere as the executives sip cognac and agree that what most Americans want is for gas to remain cheap and plentiful. "They love their cars and they love to drive," one of them says with a chuckle.

A decade after the Deepwater disaster, when lockdowns and travel bans during the coronavirus pandemic caused the global demand for oil to plunge, some analysts speculated that this ravenous need for cheap energy might finally be changing. Millions of Americans suddenly stopped taking flights and driving to work. At one point, the pandemic caused the price of oil futures to fall below zero, prompting some to suggest that fossil fuels might soon lose their significance in the metabolism of the modern world. As sea levels rose, forests burned, and the effects of climate change became increasingly dire, oil would inevitably be phased out, some argued, giving way to a new era of clean, renewable energy. A few energy companies themselves seemed to be betting on this (and making plans to cash in), including BP, which, in 2020, announced that it was investing billions in renewable energy and embarking on a path to becoming a "net zero" emissions company.

To the extent that the pandemic altered behavior in a lasting way, leading more and more people in America and other countries to work from home and travel less, it was certainly possible to imagine the age of oil ending. On the other hand, plenty of analysts had wrongly predicted this in the past. At the start of the pandemic, fossil fuels still dominated global energy consumption. The plunge in oil prices wouldn't necessarily help motivate consumers to install solar panels in their homes and telecommute. It could have the opposite effect. "If you look at SUV sales they've gone beyond their stronghold in the US," an industry analyst told the *Financial Times* a few months after the coronavirus pandemic began. "Cheap oil is likely to exacerbate that trend." Without support from the world's leading economies, moreover, the shift to cleaner forms of renewable energy was likely to remain a pipe dream. During Donald Trump's presidency, such support was sorely lacking from the United States, which, in 2017, pulled out of the Paris Agreement, the international treaty on climate change that went into effect in 2016. The move was consistent with Trump's "America First" energy policy, which called for opening 90 percent of America's coastal waters to offshore drilling. The agenda appeared to shift dramatically under Trump's successor, Joseph Biden, who, on his first day in office, announced that the United States was reentering the Paris Agreement and that he was elevating the climate crisis to a national security priority. But bold steps to address the effects of climate change were still likely to meet with stiff resistance in Washington, not only from lobbyists for the fossil fuel industry but also from many elected officials. In the Republican Party, *acknowledging* that these effects resulted from human activity was increasingly rare.

For years after the Deepwater blowout, Sara had avidly followed such developments. By the time she enrolled in film school, she'd taken to tuning them out. What she wanted now was to focus on her own well-being and to channel her feelings into the one thing that had always come naturally to her—visual art—this time by making a film that would fill in some of what *The Great Invisible* left out. When Sara watched the film, she felt validated, she told me. But she was also disappointed. One reason for the disappointment was that none of the friends she'd invited to a screening in Houston—people from Katy who, like her, had grown up in oil families—showed up. Sara attributed their absence to the fact that, like the people in Frankfurt with whom Everett Hughes met after World War II, they lacked the will to know about things that might make them feel implicated. "They didn't want to talk about it," she said. "They *still* don't want to talk about it." One guy she knew, a friend whose father had been involved in designing the well that caused the Deepwater blowout, told her that she and Stephen "should just get over it," she told me. Later, the friend apologized and told her he felt awful about what had happened. "He carried a lot of guilt," she said.

The other reason watching *The Great Invisible* upset Sara was that although she and the wife of another rig worker appeared in the film, their role was mainly limited to talking about the ordeals their husbands had undergone. As always, it seemed to her, the spouses and family members were pushed to the margins; they remained invisible. In her *Survivors* paintings, Sara had made a point of foregrounding the experiences of family members by including people like Chris Jones, whose brother died on the rig, and Natalie Roshto, who lost her husband. "I think 'survivors' extends much further than the guys that were just on that rig," she explained. "I wanted to make sure that people saw the layered effects that occur. Your children get affected; your family gets affected. It's not just—there's an explosion, your husband's hurt, that's it. It keeps extending out quite a bit."

What Sara said about rig workers was true of all forms of dirty work, it dawned on me later. It didn't just stain and tarnish the lives of individual workers. It tarnished whole families and communities, lingering in the minds and memories of the people with whom dirty workers interacted and shared their lives. The dirty work of caging human beings in crowded, violent prisons affected not only corrections officers but also their spouses and children. The dirty work of staring at screens on which human beings were blown to pieces by Hellfire missiles made it difficult for some drone operators to feel much when learning that members of their own extended families had died.

The next day, Sara invited me to watch another film, a narrative work she had made that elaborated on this theme. It opens with a U.S. Coast

Guard transmission at 9:51 p.m. on April 20, 2010, indicating that an explosion had taken place on an oil rig in the Gulf. The rest of the film toggles back and forth between two women—one in California, the other, modeled on Sara, in Texas—who learn the news and try to figure out if their husbands have survived. We watch the women register the initial shock. We see them strain not to panic as images of the burning rig appear on television. Eventually, the woman in Texas starts randomly calling the burn units of hospitals in the Gulf, and later, after seeing a report on the news describing the accident as catastrophic, stumbles onto the front patio of her house, screaming and sobbing.

The screening took place on the UCLA campus, in a theater located next to a sculpture garden dotted with jacaranda blooms. Sara and her peers gathered there to present their final projects in the narrative-film class she was taking. After her film was shown, Sara sat on the edge of the stage, fielding questions from her peers. Then she went home to "numb out," she told me. The hardest thing about watching the film wasn't looking at the images, she said at a diner where we met for brunch the following morning. It was listening to the sounds, which brought back the terror and helplessness she'd felt on the morning of the disaster. "The screams were hard," she said. "That was really hard."

After finishing brunch, Sara and I took a walk, strolling along Melrose Avenue before turning onto the street where she lived, which was lined with flower gardens and lemon trees. Her apartment was halfway down the block, a one-bedroom flat with walls adorned with Peruvian rugs and a couple of paintings. Sara had not put any of her own work up, but she did have it with her. At one point, she went into the bedroom and returned with a stack of canvases wrapped in brown paper that she laid out on the living room couch. It was the *Survivors* paintings. We stood over the portrait of Chris Jones—lips sealed, face blue, eyes brimming with rage. "He was furious," she said, "but so sad." Next to the painting of Jones was the painting of Stephen, wearing a vacant, glassy-eyed stare. Sara told me it was the hardest painting in the series to draw, because its subject was both so close to her and so distant. After the blowout, "he just took off," she said, drifting away to escape a world he'd lost faith in. Despite their separation, Sara clearly didn't blame Stephen for this. "I just think about what it's like to have such a tender soul and then have, like, everything bad put in front of you," she said.

Before I left, Sara brought out one other painting, an unfinished portrait of a young boy crouching next to a stone walkway shrouded in plants. Frogs and lizards scurry through the yard where the boy has bent down to play. It, too, was a portrait of Stephen, as he'd appeared in a family photo that his mother had sent to Sara. It showed him in more innocent times, she told me, in the one place that had always managed to bring him a measure of solace—nature. "That's where he went to," she said.

9

Dirty Tech

The stigmatized institutions where dirty work unfolds tend to be located in isolated areas with a high concentration of poor people and people of color. But aren't plenty of morally distasteful jobs performed in wealthy pockets of the country as well—places like Wall Street and Silicon Valley? And aren't many of the white-collar professionals who do these jobs also at risk of feeling dirtied by what they do?

In the spring of 2016, an assistant professor of mathematics named Jack Poulson left Stanford University for a job that seemed to carry no such risk. The job was at Google, which, that year, was ranked as the best place in America to work by *Fortune* magazine, both because of the perks—high salaries, gourmet meals served in the cafeteria of Google's lavish headquarters in Mountain View, California—and because of the company's moral cachet. "Don't be evil" was Google's motto, a slogan that would have seemed grossly out of place at a fossil fuel company. At the time, it did not seem out of place at Google, certainly not to the chorus of evangelists who hailed the digital revolution as a utopian development that would empower ordinary people and make the world a better place. The latter phrase appeared more than once in the letter that Larry Page and Sergey Brin, Google's founders, drafted when the company went public in 2004, affirming their commitment to making information more accessible and to creating a more connected world—which, it went without saying, would be a better world. "Google is not a conventional company," the letter proclaimed. "We aspire to make Google an institution that makes the world a better place."

Jack was thirty and was hired as a research scientist in Google's Artificial Intelligence Division. Many Google employees came to the company with advanced degrees in computer science. Jack's PhD was in applied math, but he possessed the technological skills and the slightly nerdy habits of a typical "smart creative," the term that Eric Schmidt, Google's executive chairman, and Jonathan Rosenberg, the former vice president of products, used to describe the ideal Google employee in their 2014 book, *How Google Works*. On weekends, Jack liked to read math textbooks and write open-source software. One of his heroes was the British philosopher and logician Bertrand Russell. He also admired iconoclastic thinkers such as George Orwell and Christopher Hitchens, writers who appealed to his skeptical sensibility and independent streak. It was this sensibility that fueled another hobby he'd developed—reading investigative journalism.

In August 2018, a piece of investigative journalism posted on an internal Google message board caught Jack's eye. The subject of the article was a version of Google search that the company was planning to launch in China. In theory, such an undertaking was perfectly in keeping with Google's stated mission to broaden access to information and make the world a better place. Repressive regimes such as China were indeed where the internet could foster the most dramatic change, many analysts had assumed at the outset of the digital revolution, foiling the ability of governments to censor and control the flow of information. But according to the article, which was written by Ryan Gallagher, a reporter at *The Intercept*, and drew on documents marked "Google confidential," Google's project in China, code-named Dragonfly, posed no threat to China's authoritarian government. Instead of challenging China's restrictions, the documents indicated, Google had chosen to comply with them, developing an Android app that would automatically identify websites barred by the Great Firewall, an online censorship apparatus that the Chinese government had erected and designed. "When a person carries out a search, banned websites will be removed from the first page of results," Gallagher reported. The app also blacklisted "sensitive queries," for which no results would appear at all. Among these blacklisted terms were "human rights" and "democracy."

Like most Google employees, Jack had never heard of the Dragonfly project. Although he wasn't involved in designing the app it described, one of his responsibilities was to improve the accuracy of a component of Google search across a variety of languages. The idea that Google customers in China would be steered only to websites vetted and approved by one of the world's most repressive governments was chilling to him. Equally chilling was the thought that Google's app, which required users to sign in to perform searches—activity that might then be linked to their personal phone numbers—could expose users to harm. The Chinese government didn't just censor content on the internet. It also conducted aggressive online surveillance, monitoring the views expressed on social media and using mobile phones to eavesdrop on the conversations of human rights activists, some of whom wound up in prison. This was hardly a secret to Google, which had entered the Chinese market in 2006. Four years later, in 2010, the company pulled out of the country, after discovering that Chinese security forces had hacked into Google and accessed the Gmail accounts of numerous dissidents. Among the targets of the operation was the artist Ai Weiwei. In an interview with *Der Spiegel*, Sergey Brin, whose family had fled the former Soviet Union during his childhood, cited his family's experience in a totalitarian country as a factor in the decision to withdraw. "At some point you have to stand back and challenge this and say, this goes beyond the line of what we're comfortable with and adopt that for moral reasons," he said.

After the *Intercept* story appeared, a coalition of human rights organizations, including Amnesty International and Human Rights in China, sent an open letter to Sundar Pichai, Google's CEO, calling on the company to cancel Dragonfly or risk "directly contributing to" human rights violations. Some Google employees dismissed this concern, arguing that Google's presence in China could help change things for the better. In Jack's view, the problem extended beyond China. He failed to see how the Dragonfly project could be reconciled with Google's AI principles, which stated that the company would neither build nor design "technologies whose purpose contravenes widely accepted principles of international law and human rights." As an email he sent to his manager made clear, he also failed to see how he could continue to do his job without feeling complicit in human rights violations. In the email, Jack laid out his concerns and indicated that unless someone with more knowledge of the situation could address them, he would resign.

"VOICE" AND "EXIT"

If fossil fuels powered the global economy, digital communication linked it together. The transmission of information across wireless networks was no less important in the metabolism of global capitalism than oil, creating an interconnected world in which products, services, and ideas could be posted, shared, and downloaded by anyone with a mobile device or laptop. For a while at least, there seemed to be few downsides to this development and nothing but praise to lavish on the forward-thinking entrepreneurs who facilitated it. "In the public consciousness, high tech is the antithesis of that old-fashioned, fossil fuel–driven industry," observed two reporters back in 2000. "The news media normally discuss the new technologies as digitally clean, trafficking in information rather than goods, thriving on creativity rather than muscle."

By the time Jack Poulson learned about the Dragonfly project, a darker view had emerged. The founders of tech companies like Facebook were no less driven by greed than the CEOs of BP and Transocean, it turned out. By 2016, it was also apparent that trafficking in information could have some very serious downsides. After Brexit and the election of Donald Trump, Facebook came under fire for its role in spreading a torrent of virulent propaganda, some of it disseminated by the consulting firm Cambridge Analytica, which gained access to the private data of millions of Facebook users who were bombarded with fake news and conspiracy theories designed to alter their voting habits. (In 2019, the Federal Trade Commission fined Facebook five billion dollars for mishandling users' personal data.) In Myanmar, Facebook served as the main conduit for incendiary messages about the Rohingya, a Muslim minority whose members were subjected to rape and killing in what the UN described as a "textbook example of ethnic cleansing."

The internet could be used not only to connect and empower people but also to surveil and manipulate them, it was becoming clear, a danger hardly limited to the political arena. According to Shoshana Zuboff, a professor emerita at Harvard Business School, the web's seemingly magical ability to anticipate users' needs masked a far more sinister development: the unchecked power that Google, Facebook, and other technology companies possessed to gather and store personal data about their customers, information accrued through hidden tracking mechanisms that was harvested to benefit targeted advertisers and modify human behavior. "Under this new regime, the precise moment at which our needs are met is also the precise moment at which our lives are plundered for behavioral data, and all for the sake of others' gain," argued Zuboff in her bracing book, *The Age of Surveillance Capitalism.* The backlash against the billion-dollar companies (Amazon, Google, Facebook) profiting from this regime was exacerbated by the sense that we were all complicit in their rise. Even as people railed at the deleterious effects of Facebook and Twitter, they grew more and more addicted to devices and screens, not infrequently venting their dismay in text messages to friends or posts on social media.

For two decades, technology companies had attracted talented young people because such companies enabled them both to make a lot of money and to feel good about the impact they were having on the world. Now some tech workers began to ask themselves—and, on occasion, their bosses—hard questions about whether the products and services they were designing might be harming the world. At Salesforce, a cloud computing company based in San Francisco, employees circulated a petition urging the company's CEO to end multiple contracts with the U.S. Customs and Border Protection agency, which they feared implicated them in the Trump administration's policy of separating parents from children at the border. At Amazon, workers protested the sale of facial recognition software to law enforcement agencies, out of concern that the technology could be used to track civil rights activists and critics of police brutality (in 2020, the sale was halted for one year).

As consciousness of the downsides of technology rose, the smugness that people in Silicon Valley had long evinced when talking about their jobs began to give way to discomfort, even shame. But while the moral luster of working in Silicon Valley had faded, this does not mean it qualified as dirty work. One key difference is that, like bankers and other white-collar professionals, tech workers who felt compromised by what they were doing had far more flexibility to do something about it. They inhabited a starkly different world than Harriet Krzykowski, who refrained from saying anything even after she learned what happened to Darren Rainey, not because she thought silencing herself was morally acceptable, but because she needed the paycheck and knew that challenging security could endanger her life. Had Harriet confronted the guards when the horrors of the "shower treatment" first came to her attention, she might have been able to avoid feeling sullied, knowing that, once she figured out what was going on, she did what she could to stop or expose it.

When Jack Poulson emailed his manager at Google to tender his conditional resignation, he still did not know exactly what was going on. But the decision to write and send the email was itself an indication that, unlike people who worked in industrial slaughterhouses and at prisons like Dade, he felt he had some leverage with his superiors. It was an example of what the economist Albert Hirschman called "voice," a mode of protest Hirschman examined in his influential book, *Exit, Voice, and Loyalty*, which analyzed the choices available to government officials, workers, and other social actors confronted by immoral or dysfunctional behavior. One of these choices was to "kick up a fuss" from within in the hope of effecting change.

Among the dirty workers I'd met who had attempted this strategy, this hope was invariably frustrated. Kicking up a fuss at Dade triggered retaliation from the guards or, as George Mallinckrodt learned, getting fired. The immigrants I'd interviewed who worked in industrial slaughterhouses knew better than to even try to voice complaints, aware the company could easily replace them by drawing on the pool of cheap, unskilled laborers. So, too, with rig workers who refrained from complaining about safety measures, and drone operators like Heather Linebaugh who saw how the military dealt with dissenters like Chelsea Manning. But Jack's experience was different, in no small part because his skills and training—a PhD in applied math from the University of Texas at Austin, a master's degree in aerospace engineering—made him far more difficult to replace and far more cognizant of his value. Jack first got a sense of this when he was still in graduate school and decided to do an internship one summer at a lab run by the Department of Energy. In the contract he was asked to sign, Jack noticed a clause stipulating that for a full year after he left, the lab would own everything related to his work. This struck him as unfair—why should open-source software that he wrote on his own time not belong to him? he wondered—so he refused to sign it. On an email chain, Jack saw that a lawyer wrote, "How dare an intern question that—fire him!" But Jack's objection to the provision did not get him terminated. Instead, the scope of his responsibilities was narrowed so that the lab could only lay claim to work he did directly for its benefit.

A few years later, after he'd published some academic papers and begun teaching computational science and engineering at Georgia Tech, Jack received an offer from Stanford. After negotiating terms that included a job for his partner, a neuroscientist, Jack accepted the offer. But on the way to Palo Alto, he learned that Stanford had reneged on this provision. He immediately called the university. "Well, okay, then I don't accept your offer," he told an administrator over the phone. Stanford promptly reversed course, arranging for his partner to work as a manager in a neuroscience lab.

From these experiences, Jack inferred something that would have been unthinkable to the dirty workers I'd met and interviewed, which was that he didn't have to accept terms of employment that he found objectionable and that using his voice could benefit him. At Google, this belief was reinforced when, about a year and a half after he was hired, he informed the company that he was moving to Toronto, where his partner had been accepted into a PhD program. In response, Google proposed to keep him at the same job in his new location but at a 40 percent lower salary, which his manager presented as a cost-of-living adjustment. The offer did not please Jack, who pointed out that while housing expenses were lower in Canada, taxes were higher. If these were the terms, he would resign, he told his manager. Later that day, one of the managers on his team showed up at his desk with a new offer, which included five hundred thousand dollars in stock options spread over four years. "Will this account for the gap?" the manager asked. Jack looked over the revised proposal and said that it would.

Jack's habit of speaking up for himself on such occasions was, in part, a reflection of his personality; a more timid, less independent-minded person might have acted differently. But it was also a reflection of the elevated stature and authority that tech workers like him possessed, not only when it came to negotiating their salaries but also over matters of ethics and conscience. After Dragonfly's existence came to light, Jack was hardly the only Google employee to exercise voice. Two weeks after *The Intercept* published its story, more than one thousand Google workers signed a letter expressing their dismay about the plan and demanding more of a say about projects the company pursued. "We urgently need more transparency, a seat at the table, and a commitment to clear and open processes," they wrote.

Unlike dirty workers in lower-skilled professions, the employees who signed this letter felt entitled to a seat at the table. Many had degrees from elite universities that inculcated graduates with a belief that their voices mattered. Tech workers often received the same message from their employers, nowhere more so than at Google, where "smart creatives" were encouraged to pose challenging questions at weekly gatherings known as TGIF meetings. (The motto "Don't be evil" was just another way "to empower employees," observed Eric Schmidt and Jonathan Rosenberg in *How Google Works*.) As it turns out, the input from workers was less welcome when it came to Dragonfly, where a TGIF meeting ended up going badly awry, leaving many employees convinced the company didn't actually value their voices.

By the time the meeting occurred, Jack realized the company had little interest in the views of its employees. The email he sent to his manager about the Dragonfly project had elicited a vague, unsatisfying reply. Frustrated by what he perceived as stonewalling, he eventually posted his concerns on an internal company message board, in a letter decrying Dragonfly as a "forfeiture of our values." This finally did get some attention from his superiors, who invited him to air his concerns at a meeting. The meeting did not go as Jack hoped it would. According to Jack, instead of clarifying what protections would exist for human rights activists in China, Jeff Dean, the head of artificial intelligence at Google, downplayed this concern, reminding Jack that the U.S. government also conducted electronic surveillance through the Foreign Intelligence Surveillance Act. When Jack brought up the letter that Amnesty International and other human rights groups had sent to Sundar Pichai, he was told that outsiders were in no position to tell Google how to run its business. At a certain point during the meeting, Jack zoned out, realizing there were limits to the leverage he had. But while this was disappointing, it did not lead him to feel trapped in the way Harriet Krzykowski had at Dade. Instead, it led him to pursue the other mode of protest that Albert Hirschman outlined in his classic study: "exit." The day after the meeting, he left the company.

In theory, the option to exit is available to any person working a dirty and demeaning job. In reality, it is a far easier option to exercise if you have the skills and education to pursue other alternatives, something most dirty workers sorely lack. The lack of alternatives is precisely what led high school graduates from "rural ghettos" to apply to work as prison guards, a "job of last resort" that few other people wanted. It is what led undocumented immigrants to work at slaughterhouses that struggled to find enough native-born Americans to hire.

What these workers labored under was the pressure of economic necessity, the same force that had compelled many of them to take on dirty jobs in the first place. Tech workers were not entirely immune to this pressure. Although many earned good salaries, living in Silicon Valley was expensive, all the more so if you had a family to support. One reason more of his peers did not resign over the Dragonfly project was that they had kids, Jack told me when we met for lunch in Toronto, about a year after he'd left Google. Although Jack did not have kids, he, too, worried about the financial consequences of losing his job—with good reason, it turned out. The year after he quit, his annual income fell by 80 percent, he told me. On the other hand, the money he'd earned while at Google (including the stock options he'd received) afforded him a measure of financial security that the dirty workers I'd met could scarcely have dreamed about. Unlike these workers, moreover, Jack had the skills and credentials to pursue alternatives, a fact underscored when, a few weeks after he left Google, an article about his resignation appeared. It was written by Ryan Gallagher, the same reporter at *The Intercept* who had broken the original story about the Dragonfly project. "There are serious worldwide repercussions to this," said Jack in the article.

Having never spoken to the media before, Jack was understandably nervous about what the personal repercussions of doing so might be, not least because, when he left Google, the company had warned him not to talk to the press. In much of Silicon Valley, he soon discovered, the response was to offer him a job. "I think I got, like, thirty companies reaching out in a forty-eight-hour period, trying to hire me—*at least* thirty," he told me. One company "flat out offered to pay me more than whatever Google paid me," he recalled. "Basically, every Silicon Valley company, sometimes multiple people from the same company, were reaching out." Offers also came to return to Stanford.

As the offers suggested, Jack did not emerge from the controversy over Dragonfly with diminished career prospects. If anything, these prospects were enhanced, burnished by his stature as a talented knowledge worker with a moral backbone, which, amid the broader backlash against Silicon Valley, made him a desirable person to hire. It also made him a desirable speaker and interview subject. After the *Intercept* article appeared, he was flooded with requests for interviews from other media outlets—Bloomberg, CNN International, Fox. He was also invited to speak at venues such as the Geneva Academy of International Humanitarian Law and Human Rights and to submit testimony at congressional hearings on the tech industry.

"I'm doing just fine," Jack said cheerfully at the restaurant in Toronto where we met for lunch, sounding not at all like a person who missed working in Silicon Valley and even less like the dirty workers I'd gotten to know. The contrast in their demeanor was as striking as the difference in their financial prospects. After leaving Google, Jack had experienced his share of challenges, struggling to find a high-tech job that would not make him feel compromised in some *other* way. (At one point, he did some consulting for a company that was helping humanitarian organizations respond to natural disasters, he told me, only to learn that one of its clients—the Department of Homeland Security—could potentially use the same technology to surveil migrants at the border, prompting him to quit.) But unlike the dirty workers I'd met, he had come away from the Dragonfly controversy relatively unscathed, with his integrity untarnished and with no trace of the moral and emotional burdens—nightmares, hair loss, guilt, shame—they endured.

Not long after meeting Jack Poulson, I spoke with another former Google employee, Laura Nolan. A graduate of Trinity College who lived in Dublin, Laura started working for Google in 2013 as a site reliability engineer, a specialized discipline that involved improving the performance and efficiency of large online software systems and services.

For several years, Laura flourished at the company, earning excellent performance reviews and taking pride in her job. In October 2017, while visiting the Bay Area, she was briefed about a new project that would improve Google's machine learning analysis of classified imagery and data. "Why are we doing this?" Laura asked out of curiosity. A colleague pulled her aside and told her it was for Project Maven, an artificial intelligence contract with the U.S. Department of Defense, the aim of which was to enhance the Pentagon's ability to track and identify objects—including vehicles and human beings—in aerial drone footage.

When Laura learned this, she was stunned. Scanning all of the world's books, reaching a billion users in Africa: this was Google's mission, she'd thought. She did not think that helping the U.S. military conduct extrajudicial killings was part of the mission. As she would subsequently learn, Project Maven was not a weapons project. It was a surveillance program designed to automate and accelerate the process of sifting through the vast volume of footage that drones recorded as they hovered over distant war zones. The distinction failed to comfort Laura, particularly as she read more about the nature of drone warfare. Among the sources she found was *Kill Chain*, a book by the investigative reporter Andrew Cockburn. The "kill chain" began with surveillance and often ended with "signature strikes" that killed innocent civilians, Cockburn argued, the same conclusion that Christopher Aaron and Heather Linebaugh had reached. Automating this process would inevitably lead to *more* surveillance, Laura feared—and, in turn, to a "scaling up" of lethal strikes.

Although Google's contract with the Pentagon was for just fifteen million dollars, it was a pilot project that could pave the way for more lucrative collaborations with the Pentagon, including a ten-billion-dollar Joint Enterprise Defense Infrastructure, Laura learned. When she shared her concerns about Project Maven with one of her directors in Dublin, he was frank about the stakes, telling her, "We have to do this because of shareholder value." So much for Google's idealism, Laura thought to herself as she began to mull leaving her job, which soon became a source of guilt rather than pride. After learning about Project Maven, she had trouble sleeping. She put on weight. She felt anxious and refrained from telling even her partner and closest friends what was wrong. The silence was obligatory, a consequence of the nondisclosure agreement she had signed, which forbade her to talk about Project Maven with anyone outside Google.

Even inside Google, few employees knew about Project Maven. This changed in February 2018 when a group of engineers who'd been working on the project aired their concerns about it on an internal message board. A week later, word of Project Maven leaked to the press. When this happened, Laura was thrilled. Finally, she could break her silence. Along with more than three thousand other Google workers, she signed a petition addressed to Google's CEO, Sundar Pichai, that called for Project Maven to be canceled. "We believe that Google should not be in the business of war," stated the petition, which was obtained and published by *The New York Times*.

The pushback from employees and heightened attention from the media eventually prompted Google to reconsider its priorities. In June 2018, Diane Greene, then the head of Google Cloud, informed the company's employees that Google would not renew the Project Maven contract when it expired. A week later, Google unveiled a new set of AI principles, affirming that it would not pursue any surveillance projects that fell outside "internationally accepted norms." On paper, this sounded like a principled position. To Laura, it sounded like a sly evasion, not least because there were no "internationally accepted norms" for surveillance projects. Far from signaling a repudiation of Project Maven, Google's new AI principles left the door open for similar work in the future, Laura was convinced. Rather than wait for this to happen, she decided to exercise the same option that Jack Poulson did: she resigned, one of roughly twenty Google employees who left the company because of Project Maven. On her last day at the office, she cried, she told me. But she also felt a weight lift, having concluded that she could no longer work as a site engineer at Google and sleep well at night.

"SPOILED IDENTITY"

As I noted at the outset of this book, dirty work has a number of essential features. One of them is that it causes substantial harm to other people or to the natural world. Another is that it causes harm to the workers *themselves*, either by leading people to feel they have betrayed their own core values or by making them feel stigmatized and devalued by others. As Laura Nolan's experience showed, holding a high-skill, high-paying white-collar job doesn't necessarily prevent people from feeling they are betraying their core values. Had she continued working at Google, her guilt would have intensified, she told me. Eventually, it might have led her to experience a moral injury. The fact that, like Jack Poulson, she *didn't* continue to work at Google showed, again, why it is so much easier for people in high-skill white-collar professions to avoid sustaining such wounds.

But what about feeling stigmatized and devalued by others? Are white-collar workers more likely to be shielded from this as well? Not according to Thomas Roulet, a lecturer at the University of Cambridge Business School. In 2015, Roulet published an article in which he argued that such a fate befell an entire white-collar profession—the banking industry—after the 2008 financial crash. To build his case, Roulet drew on an influential work of social theory, Erving Goffman's *Stigma: Notes on the Management of Spoiled Identity*. Published in 1963, Goffman's book defined stigma as an "attribute that is deeply discrediting," so much so that it "disqualified [a person] from full social acceptance." The discrediting attribute could be a bodily sign. It could also be a character trait or an affiliation with a discredited racial or religious group—any "signifier" that veered from societal norms and led an individual to be placed in a separate and defiled category, "reduced in our minds from a whole and usual person to a tainted, discounted one."

Goffman applied this theory to the "moral careers" of individuals. But as Roulet noted, it could also be applied to organizations. "Like individuals, organizations are also subject to disqualification from full social acceptance," he observed. After the 2008 financial meltdown, he argued, the financial industry qualified as such an organization, owing to the fact that the "dominant logic" of the profession—the theory of shareholder maximization, which held that a firm's sole duty was to enrich its shareholders—came to be seen as incongruous with the broader norms of society and the common good. In the shadow of the subprime crisis, what investment bankers understood to be their jobs was depicted as reprehensible. The stigma that resulted from this disjuncture was diffused through the media, Roulet maintained, which pilloried bankers for their greed. Articles such as "What Good Is Wall Street?," by the *New Yorker* journalist John Cassidy, suggested the world might be better off without firms like Goldman Sachs and Morgan Stanley.

That some media outlets relished depicting Wall Street in a negative light after the subprime meltdown is true enough. For members of a stigmatized organization, though, the bankers in Roulet's study were doing strikingly well for themselves. In his study of stigma, Goffman took it as given that an individual with a "spoiled identity" would be hobbled by this designation: "We effectively, if often unthinkingly, reduce his life chances." In the aftermath of the 2008 financial meltdown, however, the life chances of Wall Street bankers showed no signs of suffering. In the year after the crash, the average pay at Goldman Sachs, Morgan Stanley, and JPMorgan Chase *rose* 27 percent. Negative press coverage did not stop the leading firms on Wall Street from handing out tens of billions of dollars in bonuses. Nor did it damage the careers of the attorneys and lobbyists who did their bidding, lawyers like Eugene Scalia, who earned millions while helping the financial industry try to gut the Dodd-Frank Act and other regulations.

How disqualified from social acceptance could an industry that handed out such bonuses be said to be? How tainted and discounted were its employees and enablers likely to feel? Not very, common sense suggested, owing to a factor that was conspicuously absent from Roulet's analysis of the financial industry and that was largely missing from Goffman's study as well. This factor was power. As Goffman observed, stigma was acquired through relationships, social interactions during which stereotypes were formulated that led individuals to feel tainted and discounted. But as the scholars Bruce Link and Jo Phelan have argued, the potency of these stereotypes was entirely dependent on how powerful the people making them were. To illustrate this point, they described a hypothetical scenario in which patients at a mental health hospital applied derogatory labels to staff members who were arrogant and cold. The patients might mock these staff members behind their backs. They might wish to cast them into a discredited category. Even so, Link and Phelan maintained, "the staff would not end up being a stigmatized group. The patients simply do not possess the social, cultural, economic and political power to imbue their cognitions about staff with serious discriminatory consequences." It was power—including the power "to control access to major life domains like educational institutions, jobs, housing, and health care"—that put "consequential teeth" into a stigma. It was, in turn, the *absence* of power that made people vulnerable to being stigmatized, saddled with a "spoiled identity" that hampered their life chances.

Power did not shield bankers and other successful white-collar professionals (lawyers, lobbyists, tech workers) from moral opprobrium. But it made this opprobrium far less blighting and damaging: to their income, to their status, to their dignity and self-esteem. Bankers who continued earning lavish bonuses after the financial meltdown could "manage" stigma in ways that dirty workers could not—by, for example, donating money to a philanthropic organization, a virtue-affirming gesture unavailable to workers of lesser means. Even if some people looked askance at what they did, success could also breed an air of superiority and entitlement that made criticism far easier to dismiss. This helped explain why, in the aftermath of the 2008 financial crash, many investment bankers felt not tainted and discounted but indignant and aggrieved, outraged at the mere thought of having their industry regulated and of being subjected to public reproach. The outrage was an example of what the political philosopher Michael Sandel has called "meritocratic hubris," the inflated self-regard accrued by elites who managed to obtain degrees from the best law schools, business schools, and engineering programs, ostensibly because of their talent and hard work. This was the premise of meritocracy, a system that sorted people into different income brackets and career paths based on their ability to secure coveted spots at elite educational institutions. As Sandel has observed, one consequence of this system has been to diminish the dignity and self-esteem of working-class people who do not have degrees from top universities and whose fortunes have declined or stagnated in recent decades. Another is to burnish the moral credentials of society's "winners," hyper-educated achievers who have been encouraged "to regard their success as their own doing, a measure of their virtue—and to look down upon the less fortunate."

The hubris of successful meritocrats is unwarranted, some would argue, because hyper-educated achievers so often come from privileged families and wealthy backgrounds. Yet it is rooted in something successful meritocrats accurately sense, which is that even people who hold them in contempt simultaneously envy and admire them. For a job to qualify as dirty work, it needs to involve doing something that "good people"—the so-called respectable members of society—see as morally sullying and would never want to do themselves. This is true of the work performed by slaughterhouse laborers and prison guards; by the "joystick warriors" in the military; by roustabouts like Stephen Stone. It is not true of software engineers and site reliability engineers who work in Silicon Valley, or for that matter financiers and bankers on Wall Street.

"SEE NO EVIL, SPEAK NO EVIL"

Investment bankers and software engineers tend to be spared the indignities that encumber dirty workers. But this hardly means the companies they work for play no role in profiting from, and shaping, dirty work. After I spoke with Laura Nolan, it occurred to me that Project Maven could be viewed another way: as the template for the kind of dirty work that will proliferate in the future, a brave new world in which ethically troubling tasks will increasingly be delegated not to human laborers but to robots and machines. The "problem" of how to continue fighting endless wars could be solved with autonomous weapons systems programmed to strike targets on their own, obviating the need to hire "desk warriors" to do the killing. The dirty work of extracting fossil fuels could be accomplished not by hiring riggers like Stephen Stone but through cloud services and artificial intelligence. This was indeed already happening, according to a 2019 article in the online publication *Gizmodo*, which described the lucrative deals that Amazon, Google, and other tech firms had struck with oil companies to provide them with such services. "Google is using machine learning to find more oil reserves both above and below the seas, its data services are streamlining and automating extant oilfield operations, and it is helping oil companies find ways to trim costs and compete with clean energy upstarts," noted the article, which accused "Big Tech" of "automating the climate crisis."

Automating such functions would not entirely eliminate the need for human beings to participate in them: someone would still need to design and program the machines. But it could limit their participation to technical tasks for which it would be easy to diffuse responsibility, something Laura

Nolan felt was inherent to high-tech work. By way of example, Laura mentioned the hidden tracking mechanisms that companies like Google used to harvest users' personal data for advertisers, which she had come to regard as unethical. Google was the pioneer of this practice, Shoshana Zuboff argued in *The Age of Surveillance Capitalism*, honing the art of personal data mining through its "ever-expanding extraction architecture." But even if some software engineers at Google might have agreed that creating this invisible architecture was distasteful, few were likely to feel implicated in its construction and design, because their jobs involved doing other, more mundane things. "There are not that many people who are actually writing the codes that are deciding what data to collect on an individual," said Laura. "An awful lot of people are just writing code that is looking after the servers. For every one person who is doing something that could directly be seen as problematic, probably hundreds or thousands of people are just doing the housework—the plumbing and the cleaning."

"And you know, doing housework isn't wrong," she continued. "It's very easy to diffuse the responsibility for work that you might not have a direct hand in. It's like me with Project Maven. I was not being asked to write code that was tracking pine-nut farmers that were gonna get blown up. I was asked to do a thing that enabled that."

Technology alone did not ensure that responsibility would be diffused. As we have seen, it did not stop imagery analysts in the drone program from being inundated by graphic images—burned bodies, cratered homes—that could cause severe emotional distress. But the engineers at Google did not witness such images. Like Laura, they were several steps removed from the consequences of their actions, performing specialized functions whose exact purposes might not be clear even to them. As Laura noted, one reason for this was that technology was fairly easy to design for one use and then repurpose for another. While workers at a plant that manufactured tanks knew how the vehicles they were building would be used, this was not true of code, she said. "Code is a lot more plastic. In tech, you can build or design something and be told it is for purpose A and then very easily have it adapted to evil purpose B."

The plasticity enabled employers to keep workers in the dark about what was actually going on, a problem Jack Poulson noted as well. With the Dragonfly project, "employees just had no idea what the impact was of the things they were focused on," he said. "Even the privacy review team didn't know in some cases." Compounding this was another problem, which is that the work was parceled up and fragmented. While working at Google, Jack came across the work of Ursula Franklin, a physics professor who drew a distinction between holistic and prescriptive technologies. Holistic technologies were crafts performed by artisans (potters, metalsmiths) who, as Franklin put it, "control the process of their own work from beginning to end." In prescriptive technologies, by contrast, the work was divided into small steps, which led workers to focus narrowly on the discrete tasks they were given. This was the reality in the high-tech world, Jack had concluded, a world where the natural impulse was to concentrate on meeting narrow technical benchmarks that seemed devoid of moral consequences, which could easily be put out of mind. "You can very easily compartmentalize," he said.

Yet if some tech workers could be accused of benefiting from a system in which responsibility was diffused, in a different way so could all of us: the tech industry's patrons and consumers. Since the tech backlash and Silicon Valley's fall from grace, many consumers have developed a more jaundiced view of large social media and technology companies. This has notably failed to reduce the amount of time most people spend gazing at their laptops and smartphones or led the public to pay closer attention to other things, like the conditions under which the gadgets in their pockets are produced.

As numerous human rights organizations have shown, the global tech supply chain is anything but clean. On one end of this chain are the electronic gadgets on display at Best Buy and Apple Stores. On the other end are the artisanal mines featured in a report researched jointly by Amnesty International and African Resources Watch. The mines are located in Kolwezi, a city in the Democratic Republic of the Congo, which produces more than half of the world's cobalt, a key ingredient in the rechargeable ion batteries that power laptops and mobile phones (as well as electric cars). The *creuseurs* (diggers) who toil in the mines endure appalling conditions, the Amnesty report indicated, working twelve- to fourteen-hour days without gloves or face masks while breathing in toxic chemicals that can cause ailments such as the potentially fatal "hard metal lung disease." Many of the workers are children, driven to work in the mines because of desperate poverty and because their families cannot afford to send them to school. Deaths are common as creuseurs descend into narrow, makeshift mines that often collapse. (The dangerous conditions called to mind Orwell's description of the mines in *The Road to Wigan Pier*. In the metabolism of global capitalism, coal had given way to cobalt, but some things hadn't changed.) The rocks they scrape out with primitive tools are sold to companies like Congo Dongfang Mining International, a subsidiary of Huayou Cobalt, which is based in China. Eventually, some of the cobalt makes its way into products sold by companies such as Microsoft, Samsung, and Apple.

The report, "'This Is What We Die For,'" was published in 2016. Three years later, I met Catherine Mutindi, a nun who founded Bon Pasteur, a nongovernmental organization based in Kolwezi that runs a community development program offering free education to children employed in the artisanal mining sector. As the report made clear, companies like Apple and Microsoft weren't directly involved in the Congo's brutally exploitative mining sector. They relied on middlemen further "downstream"—middlemen who, in effect, did the dirty work for these companies and their customers, arranging for the cobalt to be extracted and delivered without too many questions asked. After the Amnesty International report appeared, many companies took steps to upgrade their sourcing practices, Mutindi told me, joining ventures such as the Responsible Raw Materials Initiative, which sought to improve the vetting of suppliers and to strengthen due diligence policies. According to Mutindi, however, these initiatives had done little to change conditions on the ground. Not long after we met, she sent me a link to a story in *The Guardian* about a lawsuit that fourteen Congolese families had filed against some of the world's largest tech companies, which they claimed were complicit in dangerous conditions that had caused their children to suffer serious injury or death. One of the children, paid seventy-five cents a day to haul bags of cobalt rocks, was paralyzed after falling in a tunnel. Another was buried alive when a tunnel collapsed. Among the companies named in the lawsuit were Apple, Microsoft, and Dell, targeted because they had "the authority and resources to supervise and regulate their cobalt supply chains" but failed to do so, the plaintiffs alleged. All of the companies denied this allegation. An Apple spokesperson told *The Guardian*, "Apple is deeply committed to the responsible sourcing of materials that go into our products. We've led the industry by establishing the strictest standards for our suppliers." A spokesperson for Dell said, "Dell Technologies is committed to the responsible sourcing of minerals, which includes upholding the human rights of workers at any tier of our supply chain and treating them with dignity and respect." A Microsoft spokesperson said, "If there is questionable behavior or possible violation by one of our suppliers, we investigate and take action."

The issue of child labor is graphic enough to generate headlines on occasion. A less graphic issue is corruption, which is more difficult to track but arguably even more pernicious, helping to ensure that little of the Congo's mineral wealth trickles down to its own citizens. Instead, the money ends up in the pockets of shady officials and businessmen like Dan Gertler, an Israeli billionaire whose role in the Congo was highlighted in another report I came across, by a Brussels-based NGO called Resource Matters. As the report noted, the United States sanctioned Gertler in 2017 for arranging a series of "opaque and corrupt mining and oil deals in the Democratic Republic of the Congo" (Gertler denied any involvement in corruption). These deals cost the Congo $1.3 billion, one study estimated, money that could have been used to invest in education or alternative sources of employment. Yet few technology companies had taken steps to ensure that their suppliers did not have links to Gertler. The focus of the Resource Matters report was Glencore, a Swiss company that was the world's largest producer of cobalt. According to the report, Glencore had continued paying royalties to a Gertler-affiliated entity even after he was sanctioned. The report identified fourteen companies that were "probable Glencore customers," including Apple, Samsung, BMW, and Renault. "Virtually all the companies studied in this report prohibit and condemn corruption," Resource Matters noted. But "of the 14 companies contacted, fewer than a third are prepared to recognize the corruption risk arising from the links between Gertler and Glencore," stated the report, which was titled "See No Evil, Speak No Evil: Poorly Managed Corruption Risks in the Cobalt Supply Chain." Only one company, Samsung, said it would conduct an audit to address the issue.

Since the report appeared, some initiatives have been launched to promote responsible mining practices and improve living conditions in the

Congo, including one financed by BMW, Samsung, and several other companies. But holding the tech industry accountable for troubling practices remains difficult. One reason for this is the diffusion of responsibility, this time at a corporate level, which enables so-called upstream companies selling brand-name products to distance themselves from the exploitation and corruption that takes place further downstream. The cobalt that ends up in laptops and electric cars passes through multiple hands—smelters, refiners, battery component manufacturers, battery producers—all spread across an array of countries and continents. According to one analyst I interviewed, Apple is at least four steps removed from Glencore, the company featured in the Resource Matters report. Renault is at least six steps removed, the analyst said. The jumble of intermediaries makes it easy for companies at one end of the chain to disavow responsibility for what happens at the other end.

Do the people further downstream—mid-level managerial workers whose jobs involve procuring the raw materials and components that go into laptops and smartphones, for example—think about issues such as child labor and corruption? Do they ever feel dirtied? A researcher at Global Witness, an NGO that researches the links between natural resource extraction and corruption, told me she knew of someone who worked for a mineral company based in Europe that was involved in the cobalt trade and might be willing to speak frankly about the ethical dilemmas involved. The employee was thoughtful and well-meaning, I was informed. A week or so later, I received word that the employee would indeed welcome hearing from me. The following day, I wrote an email to set up an interview.

A few days later, the employee wrote back to call the interview off, explaining that we could discuss only the business environment in the Congo, not the ethical questions that interested me. "As you are probably looking for personal stories and impressions, I fear that my contribution will be limited," the email stated. "I will therefore have to decline your invitation."

This was disappointing but not surprising. About a year earlier, I'd been put in touch with someone else who worked in the global tech supply chain and who, I was told, had wrestled thoughtfully with the ethical tensions involved. This employee worked in China, where laptops and smartphones were churned out at a high volume in factories rife with labor abuses. The further down the supply chain one went, the worse the conditions got, as manufacturers under intense pressure to supply brand-name companies with products pushed seasonal workers to the breaking point, violating overtime and child labor standards in the process. After obtaining the employee's email, I wrote to request an interview. I received no reply. When I reached out to the person who'd connected us—a professor who had studied the tech supply chain closely and knew the source personally—he attributed the silence to confidentiality agreements that the source's employer had likely pressed him to sign. "He may be (rightly) concerned about NDAs," he informed me.

NDAs—nondisclosure agreements—were ubiquitous in the high-tech world, obligating employees not to share "sensitive information" about their jobs with anyone. The agreements protected companies from the risk of having competitors steal their trade secrets. They also protected companies from embarrassment—both the companies and the other key link in the global supply chain, their customers, who might have preferred not to know about the unsavory practices that went into the wondrous gadgets they purchased. A few weeks after we met, Catherine Mutindi sent me a short video that captured these practices in real time. In the video, a Congolese soldier was standing on a muddy, rain-sodden field. Behind him were some mining trucks. At his feet was a creuseur—his chest bare, his hands tied behind his back, his work pants drenched from the mud and rainwater on the ground. The soldier standing over him was wearing rubber boots and had a gun strapped over his shoulder. In his right hand, he held another weapon, a coiled rope that he periodically lifted to flog the creuseur. As the blows fell, the creuseur rolled around on his stomach, trying to avoid the lash while sopping himself in mud. A blow grazed his head; then another landed more squarely on the back of his thigh. In the background, the voice of another man could be heard, ordering the soldier to continue whipping the creuseur in a mixture of Swahili and Chinese. Near the end of the video, a bespectacled man in khakis appeared. He was carrying a notebook and seemed to be a low-level mining official of some sort. He also seemed strikingly unperturbed, strolling by in a slow, even gait without making any effort to stop the soldier or help the creuseur.

"Difficult to believe that is in this century!" Mutindi wrote of the video. The scene indeed called to mind something out of the colonial era, when the Congo was a "Free State" run with savage cruelty by King Leopold II, who subjected African workers to extreme brutality and violence in order to extract its rubber and ivory. Yet the video was also very much of a piece with our times, capturing a "disturbing event" that remained hidden "behind the scenes of social life," invisible to the vast majority of people who filled their homes and offices with devices powered by rechargeable ion batteries. The soldier brandishing the whip and the mining official nonchalantly watching him go about his business were not the emissaries of a colonial regime. They were agents of global capitalism, doing some dirty work for the rest of us.

Epilogue

Not long after COVID-19 started spreading in the United States, stories began to appear about the anguish unfolding in the nation's hospitals, not only among the tens of thousands of patients arriving in emergency rooms short of breath, but also among beleaguered caregivers. Some of these accounts focused on problems that have long plagued the medical profession: burnout, stress, anxiety. But others examined more novel challenges, such as the ethical dilemmas facing nurses and physicians forced to decide how to apportion scarce medical resources to critically ill patients. Should the last ventilator in the ICU go to an elderly person at greater risk of dying or to a mother with two young children? Which patients should be admitted to a hospital that was already overwhelmed? Making these freighted decisions in harried, compromising circumstances could cause lasting psychological trauma, warned a June 2020 article in *Scientific American*. It could also result in moral injury. Doctors were trained "in treating one patient at a time," not in doing triage, a bioethicist quoted in the article observed. "I think the real reckoning is going to come when this is over," said a psychiatrist named Wendy Dean, who had cofounded a nonprofit to study the effects of moral injury in the health-care profession.

The moral and emotional wounds that dirty workers sustain are, like the workers themselves, unseen—"hidden injuries" that go unnoticed. Not so with medical professionals on the front lines of the coronavirus pandemic, whose travails made national news. Prominent publications not only reported on the ethical predicaments these caregivers were forced to navigate. They also invited them to tell their stories in their own words. "The collective soul of front-line health care workers is slowly and silently decaying with no rescue in sight," a physician of internal medicine wrote in *USA Today*. "We need help carrying this massive weight because we are weary."

The fact that medical professionals were given such platforms was surely not unrelated to their social prominence. It was also a product of the fact that what they did was seen as indispensable to society, even heroic. This was never more the case than during the pandemic, when physicians and first responders were hailed for putting their own health at risk to provide urgent care for others. Like home health aides and other so-called essential workers, they were performing a function that other people depended on and that was necessary to keep society running.

The idea that dirty work might also be necessary to society is more unsettling. A defining feature of such work is, after all, that it causes substantial harm, either to other people or to nonhuman animals and the environment. To a person like Toby Blomé, the Code Pink activist I visited in El Cerrito, it seems obvious enough that inflicting such harm is profoundly unnecessary. The high-tech killing conducted by drone operators happened not because targeted assassinations were essential to national security, Blomé told me, but because of the outsize influence of the military-industrial complex, a cabal of for-profit contractors and special interests that distorted America's priorities and profited from its endless wars.

But the truth is that the drone program doesn't just serve the interests of military contractors. It also serves the interests of a disengaged public that doesn't want to think too much about the endless wars being fought in its name and, thanks to the drone campaign, doesn't have to. The dirty work of conducting targeted assassinations can be left to people like Christopher Aaron and Heather Linebaugh, with few questions asked. Likewise with the dirty work of warehousing the mentally ill in jails and prisons in a society that has failed to fund mental health services. This arrangement benefits not only for-profit companies like Wexford and Corizon but also many citizens who are content to have the mentally ill disappear behind bars without dwelling on the consequences. This, too, is work essential to society, solving various "problems" that many of us would like to have taken care of, provided someone else can handle them.

Most of us don't want to hear too much about such work. We also don't want to hear too much from the people who do it on our behalf, not least because what they tell us might stir discomfort, maybe even a trace of culpability. I felt this discomfort frequently in the course of interviewing the subjects of this book. When Stephen Stone talked about people who cast judgment on the "country bumpkins" from backwoods towns who got jobs on oil rigs while giving little thought to how dependent their own lifestyles were on burning fossil fuels, I nodded in agreement. Then I drove my rental car back to the Airbnb where I was staying and felt implicated. When Harriet Krzykowski lamented the fact that many Americans saw individuals with severe mental illnesses as "throwaway people," particularly if they were poor and lacked access to treatment, I shook my head disapprovingly, but later wondered if this was, subconsciously, how I viewed them. I passed such people on the streets of New York City fairly frequently. More often than not, I felt for them, but not so deeply as to keep them in my thoughts for very long.

One reason that even people who are momentarily bothered by such thoughts avoid dwelling on them is that they feel powerless to change them. And, individually, we *are* powerless to change them. The decision of one person to buy a more fuel-efficient car (or, for that matter, an electric vehicle) will not end America's dependence on fossil fuels. Handing a couple of dollars to a mentally ill homeless person talking to himself on a street corner will not alter the fact that jails and prisons have become de facto mental health asylums. But collectively, we are not powerless to alter these things. As I noted at the outset, a core feature of dirty work is that it has a tacit mandate from "good people" who refrain from asking too many questions about it because its results do not ultimately displease them. This mandate is important. But it is not set in stone. The attitudes and assumptions that it rests upon can change, and indeed *have* changed. In the past decade, the punitive sentencing laws that fueled America's prison boom have fallen out of favor. Although warehousing mentally ill people in jails and prisons is still tolerated, this, too, may change as the backlash against mass incarceration leads us to ask questions about the social and moral costs of the status quo. Attitudes toward industrial animal agriculture have also shifted, even if, for now, this has led more often to a fixation on buying meat that is "organic" than to addressing the deplorable conditions to which workers in slaughterhouses are subjected. So, too, with our reliance on the fossil fuel industry, which more and more people are coming to realize must be phased out quickly for the planet to survive.

The belief that the troubling circumstances we have come to tolerate are impervious to change can itself become an excuse for resignation. It can fuel the kind of apathy that prevailed among the "passive democrats" that Everett Hughes described in his Frankfurt diary. The resignation is unwarranted, because, like most aspects of the social order, dirty work is not immutable. It is a function of laws and policies, of funding decisions, and of other collective choices we have made that reflect our values and priorities. Among these choices is whether to recognize the immense harm it causes, not only to innocent people and the environment but also to the people who carry it out.

The anguish such workers experience may elicit little sympathy from those who feel that anyone who participates in a cruel or violent system must be held accountable for the suffering they cause, even if, afterward, they harbor shame or regret about what they have done. As Primo Levi affirmed in "The Gray Zone," doubts and discomfort expressed after the fact by oppressors are "not enough to enroll them among the victims." But Levi also called for judgment of the low-ranking functionaries in oppressive systems to be tempered by awareness of how susceptible we all are to collaborating with power, and by an appreciation of the circumstances that lead relatively powerless people to be pushed into such roles. In contemporary America, the chief circumstance to consider, I have suggested, is inequality, which has shaped the delegation of dirty work no less than the distribution of wealth and income. More privileged Americans are spared from any involvement in such work, knowing it can be outsourced and allotted to people with fewer choices and opportunities. The result of this moral inequality is to ensure that an array of hidden injuries—stigma, shame, trauma, moral injury—are concentrated among those who are comparatively disadvantaged. These moral and emotional burdens have barely factored into the debate about inequality, perhaps because economists cannot measure and quantify them. But their effects can be equally pernicious and debilitating, shaping people's sense of self-worth, their place in the social order, and their capacity to hold on to their dignity and pride.

Inequality also shapes the geography of dirty work and who is held responsible for it. As we've seen, the blame is rarely directed at the companies that profit from it or the public officials who have passed laws and policies that perpetuate it. More typically, it falls on the least powerful people in the system, "bad apples" who are singled out after the periodic "scandals" that shock the public—the same public that spent the preceding months or years ignoring what was being done.

In fairness, one can hardly expect the public to register concern about conditions it rarely sees. Dirty work is obscured by structural invisibility: the walls and barriers that keep what happens inside prisons and industrial slaughterhouses hidden; the secrecy that envelops the drone program; the nondisclosure agreements that the middlemen overseeing the cobalt supply chain are required to sign. These arrangements have had a "civilizing" effect, pushing disturbing events "behind the scenes of social life." Yet there are limits to what even the most elaborate mechanisms of concealment can hide. In spite of the isolation and impenetrability of institutions like prisons and industrial slaughterhouses, plenty of information about what transpires within them leaks out. The secrecy cloaking the drone campaign has not stopped writers and documentary filmmakers from producing illuminating work about it, nor have nondisclosure agreements prevented NGOs from issuing detailed reports about the cobalt supply chain. The problem is not a dearth of information but the fact that many choose to avert their eyes, not only from dirty work but also from those who get stuck doing it, people with whom they almost never interact and find easy to judge.

What do we owe these workers? At minimum, it seems to me, we owe them the willingness to see them as our agents, doing work that is not disconnected from our own daily lives, and to listen to their stories, however unsettling what they tell us may be. The discomfort may run both ways, of course. The first time I met Harriet Krzykowski, she read me some excerpts from the narrative she'd written about her experience at Dade. The material was wrenching, so much so that her voice trailed off several times. Afterward, I wondered if it might turn out to be the last time we'd speak. Over the next few days, Harriet cried repeatedly during our conversations. But she also thanked me for taking the time to listen. It was intense, she said, but also "healing." The comment stayed with me, not because I felt I deserved any gratitude, but because it suggested how isolated she felt and how therapeutic the simple act of telling one's story can be.

The most effective way to help people overcome moral injury is to communalize it, Jonathan Shay argued in *Achilles in Vietnam*, providing veterans with an opportunity to share their experiences with the public. Medical professionals who risked sustaining such injuries during the coronavirus pandemic were accorded such opportunities from a public that felt indebted to them and that listened to their stories with respect and curiosity. Dirty workers like Harriet were not. As a consequence, the reckoning she engaged in was a private one, unspooling in haunting memories that she had to wrestle with alone. Missing was a parallel public reckoning, the kind of communal exercise I'd watched unfold at the VA hospital in Philadelphia, where people gathered to hear veterans talk about the moral transgressions they had committed in the course of fighting America's recent wars. Then the audience members spoke, delivering a message all dirty workers deserved to hear. "We sent you into harm's way," they chanted in unison. "We put you into situations where atrocities were possible. We share responsibility with you: for all that you have seen; for all that you have done; for all that you have failed to do."

Notes

Epigraph

vii   "The powerless must do their own dirty work": James Baldwin, *No Name in the Street* (New York: Vintage Books, 2007), 94.

Introduction

3   "There was always at least one roof": Everett C. Hughes Papers, Special Collections Research Center, University of Chicago Library. All of Hughes's quotations in the introduction come from these papers unless otherwise indicated.

4   "I am ashamed for my people": Everett Hughes, "Good People and Dirty Work," *Social Problems* 10, no. 1 (Summer 1962): 5.

5   "Having dissociated himself clearly": Ibid., 7.

6   "the most colossal and dramatic": Ibid., 3.

6   "The question concerns what is done": Ibid., 8.

8   "the moral burdens and the emotional hardships": Richard Sennett and Jonathan Cobb, *The Hidden Injuries of Class* (New York: Alfred A. Knopf, 1972), 76.

9   "None of us will ever be the same": Helen Ouyang, "I'm an E.R. Doctor in New York," *New York Times Magazine*, April 14, 2020.

13   "Dirtiness of any kind": Sigmund Freud, *Civilization and Its Discontents* (New York: W. W. Norton, 1989), 46.

13   "The distasteful is *removed behind the scenes*": Norbert Elias, *The Civilizing Process: Sociogenetic and Psychogenetic Investigations,* trans. Edmund Jephcott (1939; repr. Cambridge, Mass.: Blackwell, 1994), 121.

1. Dual Loyalties

23   "deliberate indifference": Thurgood Marshall, *Estelle v. Gamble*, Nov. 30, 1976, casetext.com/case/estelle-v-gamble.

23   "naked humans herded like cattle": Lisa Davis et al., "Deinstitutionalization? Where Have All the People Gone?," *Current Psychiatry Reports* 14, no. 3 (2012): 260.

23   "cold mercy of custodial care": Vic Digravio, "The Last Bill JFK Signed— and the Mental Health Work Still Undone," WBUR, Oct. 23, 2013, www.wbur.org/commonhealth/2013/10/23/community-mental-health-kennedy.

23 "No other affluent country": Christopher Jencks, *The Homeless* (Cambridge, Mass.: Harvard University Press, 1994), 39.

30 "to report this activity": Fred Cohen, "The Correctional Psychiatrist's Obligation to Report Patient Abuse: A Dialogue," *Correctional Mental Health Report*, Jan./Feb. 2014, 67.

30 "that their ethics": Sarah Glowa-Kollisch et al., "Data-Driven Human Rights: Using Dual Loyalty Trainings to Promote the Care of Vulnerable Patients in Jail," *Health and Human Rights Journal*, June 11, 2015, www.hhrjournal .org/2015/03/data-driven-human-rights-using-dual-loyalty-trainings-to -promote-the-care-of-vulnerable-patients-in-jail/.

31 "a deep-seated culture": "CRIPA Investigation of the New York City Department of Correction Jails on Rikers Island," U.S. Department of Justice, U.S. Attorney's Office for the Southern District of New York, Aug. 4, 2014, 3.

32 "the number of mentally ill": "The Treatment of Persons with Mental Illness in Prisons and Jails: A State Survey," Treatment Advocacy Center and National Sheriffs' Association joint report, April 8, 2014, 6.

32 "When things go wrong": Ibid., 8.

33 "There may be a limit": Marc F. Abramson, "The Criminalization of Mentally Disordered Behavior," *Hospital and Community Psychiatry* 23, no. 4 (April 1972).

33 "Perhaps the most alarming": "Treatment of Persons with Mental Illness in Prisons and Jails," 13.

34 "have been reported forcing a patient": Erving Goffman, *Asylums: Essays on the Social Situation of Mental Patients and Other Inmates* (Garden City, N.Y.: Anchor Books, 1961), 44.

38 "However distant the staff tries": Ibid., 81.

44 "Violent offenders, more often than not": Bruce Western, "Violent Offenders, Often Victims Themselves, Need More Compassion and Less Punishment," *USA Today*, Aug. 9, 2018. The findings of the Boston Reentry Study that Western and a team of scholars led are available at scholar.harvard.edu/bruce western/working-papers.

2. The Other Prisoners

47 "The evidence does not show": Katherine Fernandez Rundle, "In Custody Death Investigation Close-Out Memo." I reported on the memo's findings in "A Death in a Florida Prison Goes Unpunished," *New Yorker*, March 23, 2017, www.newyorker.com/news/daily-comment/a-death-in-a-florida-prison -goes-unpunished.

48 "We are appalled": Julie K. Brown, "Grisly Photos Stir Doubts About Darren Rainey's Death," *Miami Herald*, May 6, 2017, account.miamiherald.com /paywall/registration?resume=149026764.

48 "From time to time, we get wind": Everett Hughes, "Good People and Dirty Work," *Social Problems* 10, no. 1 (Summer 1962): 7–8.

49 "He knows quite well": Ibid., 8.

54 "small-minded, intoxicated": Dana M. Britton, *At Work in the Iron Cage: The Prison as Gendered Organization* (New York: New York University Press, 2003), 53.

54 "a last resort for the unskilled": David J. Rothman, *Conscience and Convenience: The Asylum and Its Alternatives in Progressive America* (New Brunswick, N.J.: Transaction, 2012), 146.

54 "the exercise of authority": Frank Tannenbaum, *Wall Shadows: A Study in American Prisons* (New York: G. P. Putnam's Sons, 1922), 29.

55 "For his own clear conscience's sake": Ibid., 25.

55 "The job of the guard": Gresham M. Sykes, *The Society of Captives: A Study of a Maximum Security Prison* (Princeton, N.J.: Princeton University Press, 1958), 59.

55 "a program for the recruitment": James B. Jacobs and Lawrence J. Kraft, "Integrating the Keepers: A Comparison of Black and White Prison Guards in Illinois," *Social Problems* 25, no. 3 (Feb. 1978): 304.

55 "Black inmates want black staff": Ibid.

56 "A majority of guards": Ibid., 316.

58 "Stigmatized places are more likely": John M. Eason, *Big House on the Prairie: Rise of the Rural Ghetto and Prison Proliferation* (Chicago: University of Chicago Press, 2017), 16.

59 "For after all, if we were to ask": Quoted in Britton, *At Work in the Iron Cage*, 51.

59 "service as a primary factor": Lewis Z. Schlosser, David A. Safran, and Christopher A. Sbaratta, "Reasons for Choosing a Correction Officer Career," *Psychological Services* 7, no. 1 (2010): 34.

59 "grew up dreaming": Britton, *At Work in the Iron Cage*, 80.

62 "discrepancies between their own ethical standards": Kelsey Kauffman, *Prison Officers and Their World* (Cambridge, Mass.: Harvard University Press, 1988), 223–24.

62 "sellouts": Lateshia Beachum and Brittany Shammas, "Black Officers, Torn Between Badge and Culture, Face Uniquely Painful Questions and Insults," *Washington Post*, Oct. 9, 2020, www.washingtonpost.com/history/2020/10/09/black-law-enforcement-protests/.

65 "It's time to stop": Linda Kleindiest, "Chiles Begins Campaign for 'Safe Streets' Program," *Florida Sun Sentinel*, April 20, 1993.

66 "The miserly Negro trader": Daniel Hundley, *Social Relations in Our Southern States* (New York: H. B. Price, 1860), 140.

66 "horrorstruck and disgusted": Robert Gudmestad, *A Troublesome Commerce: The Transformation of the Interstate Slave Trade* (Baton Rouge: Louisiana State University Press, 2003), 36.

66 "therapeutic": Ibid., 85.

67 "vulgar drudgery": Quoted ibid., 158.

67 "There was apparently little stigma": Walter Johnson, *Soul by Soul: Life Inside the Antebellum Slave Market* (Cambridge, Mass.: Harvard University Press, 1999), 54–55.

67 "See those gentlemen": Quoted in Gudmestad, *A Troublesome Commerce*, 157.

69 "Despite all of the information": Dylan Hadre and Emily Widre, "Failing Grades: States' Responses to COVID-19 in Jails and Prisons," Prison Policy Initiative, June 25, 2020, www.prisonpolicy.org/reports/failing_grades.html.

69 "They're terrified": Brendon Derr, Rebecca Grisbach, and Danya Issawi, "States Are Shutting Down Prisons as Guards Are Crippled by COVID-19," *New York Times*, Jan. 2, 2021.

69 "This is crazy": Melissa Montoya, "No Indictment for Corrections Officers in Inmate Death," *News-Press*, July 7, 2015, www.news-press.com/story/news /local/2015/07/07/indictment-corrections-officers-inmate-death/29848827/.

71 "I never tell anyone what I do": Quoted in Didier Fassin, *Prison Worlds: An Ethnography of the Carceral Condition* (Cambridge, U.K.: Polity Press, 2017), 146.

71 "why the image": Quoted in Fassin, *Prison Worlds*, 329.

72 "I have the best job": Jessica Benko, "The Radical Humaneness of Norway's Halden Prison," *New York Times Magazine*, March 19, 2015, www.nytimes .com/2015/03/29/magazine/the-radical-humaneness-of-norways-halden -prison.html.

3. Civilized Punishment

76 "Americans took enormous pride": David J. Rothman, "Perfecting the Prison: United States, 1789–1865," in *The Oxford History of the Prison*, ed. Norval Morris and David J. Rothman (New York: Oxford University Press, 1998), 100.

76 "Nothing was concealed or hidden": Charles Dickens, *The Works of Charles Dickens: American Notes* (London: Chapman and Hall, 1907), 116–17.

76 "unobtrusive margins": John Pratt, *Punishment and Civilization: Penal Tolerance and Intolerance in Modern Society* (Thousand Oaks, Calif.: Sage, 2002), 52.

76 "I hold this slow": Dickens, *American Notes*, 116.

77 "threshold of repugnance": Norbert Elias, *The Civilizing Process: Sociogenetic and Psychogenetic Investigations*, trans. Edmund Jephcott (1939; repr. Cambridge, Mass.: Blackwell, 1994), 121.

78 "Routine violence and suffering": David Garland, *Punishment and Modern Society: A Study in Social Theory* (Chicago: University of Chicago Press, 1990), 243.

78 "the civilized prison became the invisible prison": John Pratt, "Norbert Elias and the Civilized Prison," *British Journal of Sociology* 50, no. 2 (1999): 287.

83 "We're no longer to the point": Quoted in Regan McCarthy, "Department of Corrections Workers Share Their View from the Inside," WUSF Public Media, March 12, 2015, wusfnews.wusf.usf.edu/2015-03-12/department-of -corrections-workers-share-their-view-from-the-inside.

83 "a bad tag": "Prison Bill Emerges in the House—but Without Oversight Commission," *Palm Beach Post*, March 19, 2015, www.palmbeachpost.com/ 2015/03/19/prison-bill-emerges-in-house-but-without-oversight-commission/.

86 "We order surgery and they don't come in": Quoted in Pat Beall, "Inmate Was Getting Only Tylenol for Cancer," *Palm Beach Post*, Aug. 1, 2018,

www.palmbeachpost.com/news/inmate-was-getting-only-tylenol-for
-cancer/luLV1P4koWjXqCau46piMK/.

86 "serious problems with the delivery": Matthew Clarke, "Court's Expert
Says Medical Care at Idaho Prison Is Unconstitutional," *Prison Legal News*,
Aug. 25, 2016, www.prisonlegalnews.org/news/2016/aug/25/courts-expert
-says-medical-care-idaho-prison-unconstitutional/.

87 "every dime saved on prisoner care": Will Tucker, "Profits vs. Prisoners:
How the Largest U.S. Prison Health Care Provider Puts Lives in Danger,"
Southern Poverty Law Center, Oct. 27, 2016, www.splcenter.org/20161027
/profits-vs-prisoners-how-largest-us-prison-health-care-provider-puts-lives
-danger.

87 "Once or twice a week": Ibid.

87 "A special burden of accountability": John D. Donahue, *The Privatization
Decision: Public Ends, Private Means* (New York: Basic Books, 1989), 11.

88 "It's a decision that's best": "DOC to Move Forward with Prison Health
Privatization," News Service of Florida, July 18, 2012, www.northescambia
.com/2012/07/doc-to-move-forward-with-prison-health-privatization.

88 "The greater their social distance from us": Everett Hughes, "Good People
and Dirty Work," *Social Problems* 10, no. 1 (Summer 1962): 9.

89 "as many as 125,000 people": Supreme Court of the State of Florida,
"Mental Health: Transforming Florida's Mental Health System," 2007, 10,
www.floridasupremecourt.org/content/download/243049/file/11-14-2007
_Mental_Health_Report.pdf.

90 "nowhere to go": Liz Szabo, "Cost of Not Caring," *USA Today*, May 12, 2014,
www.usatoday.com/story/news/nation/2014/05/12/mental-health-system
-crisis/7746535.

91 "institutional brutality is deeply ingrained": Steve J. Martin, "It's Not Just
Policing That Needs Reform. Prisons Need It, Too," *Washington Post*, July 6,
2020, www.washingtonpost.com/opinions/2020/07/06/its-not-just-policing
-that-needs-reform-prisons-need-it-too.

95 "Although every defense lawyer": Julie K. Brown, "Rainey Family Settles Suit
in Prison Shower Death," *Miami Herald*, Jan. 26, 2018.

4. Joystick Warriors

102 "I am a torturer": Eric Fair, *Consequence: A Memoir* (New York: Henry Holt,
2016), 239.

102–103 "the antithesis of the dirty, intimate work": Mark Mazzetti, *The Way
of the Knife: The CIA, a Secret Army, and a War at the Ends of the Earth* (New
York: Penguin Press, 2013), 121.

103 "Every drone strike is an execution": Quoted ibid., 319.

104 "It is hard to imagine": Quoted in Nick Cumming-Bruce, "The Killing of
Qassim Suleimani Was Unlawful, Says UN Expert," *New York Times*, June
9, 2020, www.nytimes.com/2020/07/09/world/middleeast/qassim-suleimani
-killing-unlawful.htm.

105 "credible evidence of civilian casualties": "Out of the Shadows: Recommendations to Advance Transparency in the Use of Lethal Force," Columbia Law School Human Rights Clinic and Sana'a Center for Strategic Studies, June 2017, 6, web.law.columbia.edu/sites/default/files/microsites/human-rights-institute/out_of_the_shadows.pdf.

106 "are based thousands of miles away": Quoted in Charlie Savage, "U.N. Report Highly Critical of U.S. Drone Attacks," *New York Times*, June 2, 2010, www.nytimes.com/2010/06/03/world/03drones.html.

109 "I shall argue what I've come to strongly believe": Jonathan Shay, *Achilles in Vietnam: Combat Trauma and the Undoing of Character* (New York: Simon & Schuster, 1994), 20.

109 "perpetrating, failing to prevent": Brett T. Litz et al., "Moral Injury and Moral Repair in War Veterans," *Clinical Psychology Review* 29, no. 8 (2009): 695.

110 "significant, independent predictor": Shira Maguen et al., "The Impact of Killing on Mental Health Symptoms in Gulf War Veterans," *Psychological Trauma Theory Research Practice and Policy*, no. 3 (2011): 25.

111 "PTSD as a diagnosis has a tendency": Tyler Boudreau, "The Morally Injured," *Massachusetts Review* 52, no. 3–4 (2011): 750.

113 "The object of waging a war": George Orwell, *1984* (New York: Harcourt, 1977), 178.

115 "and my presence was guilt enough": Tim O'Brien, *The Things They Carried* (New York: Houghton Mifflin Harcourt, 2009), 171.

115 "Your nerves, hell": Quoted in Thomas E. Ricks, *The Generals: American Military Command from World War II to Today* (New York: Penguin, 2012), 60.

115 "a filthy and unfathomable war": Robert J. Lifton, *Home from the War: Vietnam Veterans, Neither Victims nor Executioners* (New York: Simon & Schuster, 1973), 100.

115 "about behavior required for survival": Litz et al., "Moral Injury and Moral Repair," 696.

115 "asterisks in the clinician's handbook": David J. Morris, *The Evil Hours: A Biography of Post-Traumatic Stress Disorder* (New York: Houghton Mifflin Harcourt, 2015), 204.

116 "may unknowingly provide nonverbal": Litz et al., "Moral Injury and Moral Repair," 696.

116 "You can't kill unless you are prepared": M. Shane Riza, *Killing Without Heart: Limits on Robotic Warfare in an Age of Persistent Conflict* (Washington, D.C.: Potomac Books, 2013), 55.

122 "veterans are often the ones left holding": Chris J. Antal, "Patient to Prophet: Building Adaptive Capacity in Veterans Who Suffer Moral Injury" (PhD diss., Hartford Seminary, 2017), 42.

5. The Other 1 Percent

132 "If the Civil War system": Michael Sandel, *Justice: What's the Right Thing to Do?* (New York: Farrar, Straus and Giroux, 2009), 77.

133 "Only if people have a reasonable range": Ibid., 82.

133 "70% of the volunteers in the city": Quoted ibid., 83. See also Charles Rangel, "Why I Want the Draft," *New York Daily News*, Nov. 22, 2006.

133 "fifty percent more non-fatal casualties": Douglas L. Kriner and Francis X. Shen, "Invisible Inequality: The Two Americas of Military Sacrifice," *University of Memphis Law Review* 46 (2016): 563.

133 "it matters not where you live": Thomas Paine, *The Crisis*, Dec. 23, 1776, www.ushistory.org/paine/crisis/c-01.htm.

134 "benefited the economically successful": Beth Bailey, *America's Army: Making the All-Volunteer Force* (Cambridge, Mass.: Harvard University Press, 2009), 6.

134 "all citizens have equal rights": Andrew Bacevich, *Breach of Trust: How Americans Failed Their Soldiers and Their Country* (New York: Picador, 2014), 19.

134 "Few of the very rich": Ibid., 43.

135 "A hugely preponderant": Quoted in Sandel, *Justice*, 86.

136 "We can't just bomb someone": Interviewed in *National Bird*, directed by Sonia Kennebeck (Independent Lens, 2016).

138 "This is what's coming": Quoted in Francisco Cantú, *The Line Becomes a River: Dispatches from the Border* (New York: Riverhead Books, 2018), 20.

138 "Long confused with PTSD": Quoted ibid., 150. See also David Wood, *What Have We Done: The Moral Injury of Our Longest Wars* (New York: Little, Brown, 2016).

139 Some labeled him a "Nazi": Simon Romero, "Border Patrol Memoir Ignites Dispute," *New York Times*, May 19, 2018, www.nytimes.com/2018/05/19 /us/francisco-cantu-border-patrol.html.

140 "to burden them with guilt": Primo Levi, *The Drowned and the Saved* (New York: Vintage Books, 1989), 43.

140 "the room for choices": Ibid., 50.

140 "with pity and rigor": Ibid., 60.

142 "Some of them grew up": Cantú, *The Line Becomes a River*, 24.

142 "It's a job in a county": Brittny Mejia, "Complicated Identities Wrestling with Internal Conflict," *Los Angeles Times*, April 23, 2018, enewspaper.latimes.com /infinity/article_share.aspx?guid=ac6f8add-169e-4509-93dc-77e27b5ef3f7.

143 "You're arresting my *compadres*": Quoted in Josiah McC. Heyman, "U.S. Immigration Officers of Mexican Ancestry as Mexican Americans, Citizens, and Immigration Police," *Current Anthropology* 43, no. 3 (June 2002): 487.

143 "kid killer": Quoted in Manny Fernandez et al., "'People Actively Hate Us': Inside the Border Patrol's Morale Crisis," *New York Times*, Sept. 15, 2019, www.nytimes.com/2019/09/15/us/border-patrol-culture.html.

144 "It remains true": Levi, *The Drowned and the Saved*, 49.

145 "For most Americans": Francisco Cantú, "Cages Are Cruel. The Desert Is, Too," *New York Times*, June 30, 2018.

146 "Whenever I read comments": Heather Linebaugh, "I Worked on the US Drone Program. The Public Should Know What Really Goes On," *Guardian*, Dec. 29, 2013, www.theguardian.com/profile/heather-linebaugh.

147 "to someone who is listening": Jonathan Shay, *Achilles in Vietnam: Combat Trauma and the Undoing of Character* (New York: Simon & Schuster, 1994), 4.

148 "The public simply did not have": Jameel Jaffer, *The Drone Memos: Targeted Killing, Secrecy, and the Law* (New York: New Press, 2016), 29.

148 "superfluous reporting requirements": Jennifer Gibson, "Why Trump's Actions Have Put Civilians at More Risk," Bureau of Investigative Journalism, March 7, 2019, www.thebureauinvestigates.com/stories/2019-03-07/opinion-why-trumps-actions-have-put-civilians-at-more-risk.

148 "to present warfare as a form of virtue": Samuel Moyn, "A War Without Civilian Deaths?," *New Republic*, Oct. 23, 2018, newrepublic.com/article/151560/damage-control-book-review-nick-mcdonell-bodies-person.

149 "sculpt the world around it": Trevor Paglen, *Blank Spots on the Map: The Dark Geography of the Pentagon's Secret World* (New York: Dutton, 2009), 17.

150 "Blank spots on maps outline": Ibid.

150–151 "I'm thankful that my doctors": Quoted in Jonathan S. Landay, "Obama's Drone War Kills 'Others,' Not Just Al Qaida Leaders," McClatchy News, April 9, 2013, www.mcclatchydc.com/news/nation-world/world/article24747826.html.

151 "pieces of flesh and put[ting] them in a coffin": Quoted in International Human Rights and Conflict Resolution Clinic at Stanford Law School and Global Justice Clinic at NYU School of Law, "Living Under Drones: Death, Injury and Trauma to Civilians from US Drone Practices in Pakistan" (2012), 60.

151 "The blank spots on the map": Rebecca Solnit, "The Visibility Wars," in *Invisible: Covert Operations and Classified Landscapes*, a book of Trevor Paglen's photographs and projects on secrecy and surveillance (New York: Aperture, 2010), 6.

6. Shadow People

157 "One can't live": Lawrence Wright, *God Save Texas* (New York: Alfred A. Knopf, 2018), 299–300.

161 "There are people out there": Interviewed in "Our Town—Part Two," narrated by Ira Glass and Miki Meek, *This American Life*, Dec. 8, 2017, www.thisamericanlife.org/633/our-town-part-two.

162 "Employers feel under no compulsion": Philip Martin, "The Missing Bridge: How Immigrant Networks Keep Americans out of Dirty Jobs," *Population and Environment* 14, no. 6 (1993): 539.

163 "It made us all think": Interviewed in "Our Town—Part Two."

163 "tortured flesh": Jonathan Safran Foer, *Eating Animals* (New York: Little, Brown, 2009), 143.

163 "regularly ripped off the heads": Ibid., 182.

164 "The blasé attitude towards unbearable suffering": "Shocking Investigation at a Turkey Slaughterhouse," People for the Ethical Treatment of Animals website, www.peta.org/blog/shocking-investigation-turkey-slaughterhouse/.

164 "In Tokugawa-era Japan": Wilson J. Warren, *Tied to the Great Packing Machine: The Midwest and Meatpacking* (Iowa City: University of Iowa Press, 2006), 135–36.

164 "they who delight": Quoted ibid., 135.

164 "a very river of death": Upton Sinclair, *The Jungle* (1906; repr. New York: Penguin Books, 2006), 36.

165 "The novel I plan": Quoted in Leon Harris, *Upton Sinclair: American Rebel* (New York: Thomas Y. Crowell, 1975), 78.

165 "There was not even a place": Sinclair, *The Jungle*, 114.

165 "I aimed at the public's heart": Quoted in Harris, *Upton Sinclair*, 71.

166 "social and cultural apartheid": Roger Horowitz, *"Negro and White, Unite and Fight!": A Social History of Industrial Unionism in Meatpacking, 1930–90* (Urbana: University of Illinois Press, 1997), 63.

166 "defeat prejudice": Ibid., 74.

167 "My scars are many": Quoted ibid., 245.

167 "All the abuses described": Lance Compa, "Blood, Sweat, and Fear: Workers' Rights in US Meat and Poultry Plants," Human Rights Watch, 2005, 101.

170 "causes more than bodily pain": Angela Stuesse, *Scratching Out a Living: Latinos, Race, and Work in the Deep South* (Oakland: University of California Press, 2016), 127.

172 "They say every Englishman": Rudyard Kipling, *From Sea to Sea: Letters of Travel* (1899; repr. Charleston, SC: BiblioBazaar, 2016), 2:148.

173 "with a high percentage": Kira Burkhart et al., "Water Pollution from Slaughterhouses," Environmental Integrity Project, 2018, 3.

174 "They have been *chickenized*": Christopher Leonard, *The Meat Racket: The Secret Takeover of America's Food Business* (New York: Simon & Schuster, 2014), 145.

174 "We believe if we are": "Sanderson Farms Continues Its Mission in Transparency," press release, 2018, sandersonfarms.com/press-releases/sanderson-farms-continues-mission-transparency/.

179 "Without the Mexican": David Montejano, *Anglos and Mexicans in the Making of Texas, 1836–1986* (Austin: University of Texas Press, 1987), 183.

179 "the Mexican was inferior": Ibid., 228.

179 "had to be taught": Ibid., 231.

179 "matter out of place": Mary Douglas, *Purity and Danger: An Analysis of Concepts of Pollution and Taboo* (London: Routledge, 2002), 44.

179 "Where there is dirt": Ibid., 44.

179 "polluting person": Ibid., 140.

180 "this people [Jews] should": Quoted in Jerry Muller, *Capitalism and the Jews, vol. 2, Belonging, 1492–1900* (Princeton, N.J.: Princeton University Press, 2010), 10.

180 "potentates of England's Jews": Simon Schama, *The Story of the Jews* (New York: HarperCollins, 2017), 317.

180 "sucked dry, injured": R. Po-chia Hsia and Hartmut Lehmann, eds., *In and out of the Ghetto: Jewish-Gentile Relations in Late Medieval and Early Modern Germany* (Cambridge, U.K.: Cambridge University Press, 1995), 164.

180 "the breaking of interest slavery": Muller, *Capitalism and the Jews*, 31.

181 "When a Jewish lender died": Schama, *The Story of the Jews*, 2: 315–16.

181 "The way they came": Quoted in Richard Fausset, "After ICE Raids, a

Reckoning in Mississippi's Chicken County," *New York Times*, Dec. 28, 2019, www.nytimes.com/2019/12/28/us/mississippi-ice-raids-poultry-plants.html.

7. "Essential Workers"

188 "No other country raises": Michael Pollan, *The Omnivore's Dilemma: A Natural History of Four Meals* (New York: Penguin Press, 2006), 333.

189 "not only whole fish": Norbert Elias, *The Civilizing Process: Sociogenetic and Psychogenetic Investigations, trans. Edmund Jephcott* (1939; repr. Cambridge, Mass.: Blackwell, 1994), 118.

189 "too graphic for the viewing public": Gail A. Eisnitz, *Slaughterhouse: The Shocking Story of Greed, Neglect, and Inhumane Treatment Inside the Meat Industry* (Amherst, N.Y.: Prometheus Books, 1997), 214.

189 "zone of confinement": Timothy Pachirat, *Every Twelve Seconds: Industrialized Slaughter and the Politics of Sight* (New Haven, Conn.: Yale University Press, 2013), 97.

190 "The cattle jump and kick": Ibid., 151.

190 "The worst thing": Quoted in Eisnitz, *Slaughterhouse*, 87–88.

190 "Nobody wants to do that": Quoted in Pachirat, *Every Twelve Seconds*, 151.

191 "She maintained, passionately and with conviction": Ibid., 160.

192 "Killing and cutting up the animals": Lance Compa, "Blood, Sweat, and Fear: Workers' Rights in US Meat and Poultry Plants," Human Rights Watch, 2005, 11.

192 "Every week throughout the South": Scott Bronstein, "Special Report—Chicken: How Safe?," *Atlanta Journal-Constitution*, May 26, 1991.

193 "They don't talk about it publicly": Kimberly Kindy, "At Chicken Plants, Chemicals Blamed for Health Ailments Are Poised to Proliferate," *Washington Post*, April 25, 2013.

195 "JBS was in touch": Quoted in Robert Klemko and Kimberly Kindy, "He Fled Congo to Work in a U.S. Meat Plant. Then He—and Hundreds of His Co-workers—Got the Coronavirus," *Washington Post*, Aug. 6, 2020.

195 "The food supply chain": "A Delicate Balance: Feeding the Nation and Keeping Our Employees Healthy," Tyson ad, April 27, 2020. The ad can be found at www.washingtonpost.com/context/tyson-ad/86b9290d-115b-4628-ad80-0e679dcd2669.

195 "It is important": Donald J. Trump, "Executive Order on Delegating Authority Under the DPA with Respect to Food Supply Chain Resources During the National Emergency Caused by the Outbreak of COVID-19," April 28, 2020, www.whitehouse.gov/presidential-actions/executive-order-delegating-authority-dpa-respect-food-supply-chain-resources-national-emergency-caused-outbreak-covid-19/.

196 "In all, a record amount": Michael Corkery and David Yaffe-Bellany, "As Meat Plants Stayed Open to Feed Americans, Exports to China Surged," *New York Times*, June 16, 2020, www.nytimes.com/2020/06/16/business/meat-industry-china-pork.html.

197 "a consistent relationship": Bruce P. Bernard, ed., "Musculoskeletal Disorders and Workplace Factors," U.S. Department of Health and Human Services, July 1997, iii.

198 "I haven't seen conditions": Quoted in Nancy Cleeland, "Union Decries Conditions at Pilgrim's Pride Chicken Plant," *Los Angeles Times*, Feb. 27, 2002.

200 "commercial shackles are unjust": Quoted in John Samples, "James Madison's Vision of Liberty," *Cato Policy Report* 23, no. 2 (March/April 2001): 11, www .cato.org/sites/cato.org/files/serials/files/policy-report/2001/3/madison.pdf.

202 "These are great, great people": "Trump Makes Appearance at the RNC with Frontline Workers," CNN, Aug. 25, 2020, www.cnn.com/politics/live-news /rnc-2020-day-1/h_4bb5f99b5b708420912a9e00b58ddc99.

202 "an entire region hostage": Holly Ellyatt, "German District Sees Lockdown Return as Country Tries to Suppress Regional Outbreaks," CNBC, June 23, 2020, www.cnbc.com/2020/06/23/germany-is-struggling-with -more-coronavirus-outbreaks.html.

202 "organized irresponsibility": Phillip Grull, "German Labour Minister Announces Stricter Standards in the Meat Industry," EURACTIV Germany, July 29, 2020, www.euractiv.com/section/agriculture-food/news/german-labour-minister -announces-stricter-standards-in-the-meat-industry/.

203 "*Nothing* can bring these workers": Quoted in "Death of Four Workers Prompts Deeper Look at DuPont Safety Practices," OSHA News Release— Region 6, Department of Labor, July 9, 2015, www.osha.gov/news/newsreleases /region6/07092015.

203 "absent extraordinary circumstances": Noam Scheiber, "Labor Department Curbs Announcements of Company Violations," *New York Times*, Oct. 23, 2020, www.nytimes.com/2020/10/23/business/economy/labor-department -memo.html.

204 "home and social" aspects: Liz Crampton, "Azar Blames 'Home and Social Conditions' for the Meatpacking Crisis," *Politico*, May 8, 2020, www.politico .com/newsletters/morning-agriculture/2020/05/08/azar-blames-home-and -social-conditions-for-the-meatpacking-crisis-787452.

205 "a nicety that makes sense": Quoted in Michael Grabell, Claire Perlman, and Bernice Yeung, "Emails Reveal Chaos as Meatpacking Companies Fought Health Agencies over COVID-19 Outbreaks in Their Plants," ProPublica, June 12, 2020, www.propublica.org/article/emails-reveal-chaos -as-meatpacking-companies-fought-health-agencies-over-covid-19-outbreaks -in-their-plants.

206 "What do we see": Interviewed in Christina Stella, "Immigrant Meatpackers Say They're Being Blamed for Spread of COVID-19," NPR, Aug. 10, 2020, www.npr.org/2020/08/10/900766712/immigrant-meatpackers-say-theyre -being-blamed-for-spread-of-covid-19.

207 "We live in the shadows": Quoted in Margaret Gray, "The Dark Side of Local," *Jacobin*, Aug. 21, 2016, www.jacobinmag.com/2016/08/farmworkers-local -locavore-agriculture-exploitation/.

207 "They don't eat the workers": Quoted in Margaret Gray, *Labor and the Locavore: The Making of a Comprehensive Food Ethic* (Berkeley: University of California Press, 2013), 138.

8. Dirty Energy

212 "heat, noise, confusion": George Orwell, *The Road to Wigan Pier* 1937; repr. (New York: Berkley Publishing, 1961), 21.

213 "The most definitely distinctive thing": Ibid., 36.

213 "fossil fuels still supplied": "World Energy Outlook 2019," IEA report, www .iea.org/reports/world-energy-outlook-2019/oil.

216 "serious deficiencies": National Commission on the BP Deepwater Horizon Oil Spill and Offshore Drilling," Deepwater: The Gulf Oil Disaster and the Future of Offshore Drilling, Jan. 2011, 221, www.govinfo.gov/content/pkg /GPO-OILCOMMISSION/pdf/GPO-OILCOMMISSION.pdf.

220 "a series of identifiable mistakes": Ibid., vii.

220 "indignant rage": Jonathan Shay, *Achilles in Vietnam: Combat Trauma and the Undoing of Character* (New York: Simon & Schuster, 1994), 21.

222 "A path to a life": David Barstow, David Rohde, and Stephanie Saul, "Deepwater Horizon's Final Hours," *New York Times*, Dec. 25, 2010.

224 "to get *away* from that": Quoted in William R. Freudenburg and Robert Gramling, *Oil in Troubled Waters: Perceptions, Politics, and the Battle over Offshore Drilling* (Albany: State University of New York Press, 1994), 51.

224 "The more oil, the more jobs": Arlie Russell Hochschild, *Strangers in Their Own Land: Anger and Mourning on the American Right* (New York: New Press, 2016), 73.

224 "least resistant personalities": Ibid., 80–81.

225 "It was like we had been invaded": Quoted in Jason Theriot, *American Energy, Imperiled Coast: Oil and Gas Development in Louisiana's Wetlands* (Baton Rouge: Louisiana State University Press, 2014), 43.

226 "Those leaders carefully walked": Ibid., 201.

230 "more than four times higher": National Commission on the BP Deepwater Horizon Oil Spill and Offshore Drilling, "Deepwater," 225.

230 "more as a partner": Ibid., 71–72.

231 "Help is on the way": Quoted in Eric Lipton, "Trump Rollback Targets Offshore Rules 'Written with Human Blood,'" *New York Times*, March 10, 2018, www.nytimes.com/2018/03/10/business/offshore-drilling-trump-administration .html.

236 "It makes me feel guilty": Interviewed in, *The Great Invisible*, directed by Margaret Brown (Radius-TWC, 2014).

237 "If you look at SUV sales": Quoted in David Sheppard, "Pandemic Crisis Offers Glimpse into Oil Industry's Future," *Financial Times*, May 3, 2020, www.ft.com/content/99fc40be-83aa-11ea-b872-8db45d5f6714.

9. Dirty Tech

241 "Google is not": Sergey Brin and Larry Page, "'An Owner's Manual' for Google's Shareholders," 2004 Founders' IPO Letter, abc.xyz/investor /founders-letters/2004-ipo-letter/.

242 "smart creative": Eric Schmidt and Jonathan Rosenberg, *How Google Works*, with Alan Eagle (New York: Grand Central Publishing, 2014), 17.

242 "When a person carries out": Ryan Gallagher, "Google Plans to Launch Censored Research Engine in China, Leaked Documents Reveal," *Intercept*, Aug. 1, 2018, theintercept.com/2018/08/01/google-china-search-engine -censorship/.

243 "At some point you have to stand back": Quoted in "It Was a Real Step Backward," *Spiegel International*, March 3, 2010, www.spiegel.de /international/business/google-co-founder-on-pulling-out-of-china-it-was-a -real-step-backward-a-686269.html.

243 "directly contributing to": "Open Letter: Google Must Not Capitulate on Human Rights to Gain Access to China," Aug. 28, 2010, www.amnesty .org/en/latest/news/2018/08/open-letter-to-google-on-reported-plans-to -launch-a-censored-search-engine-in-china/.

244 "In the public consciousness": Quoted in David Naguib Pellow and Lisa Sun-Hee Park, *The Silicon Valley of Dreams: Environmental Injustice, Immigrant Workers, and the High-Tech Global Economy* (New York: New York University Press, 2002), 1.

244 "textbook example of ethnic": Stephanie Nebehay, "U.N. Sees 'Textbook Example of Ethnic Cleansing' in Myanmar," Reuters, Sept. 11, 2017, www .reuters.com/article/us-myanmar-rohingya-un/u-n-sees-textbook-example -of-ethnic-cleansing-in-myanmar-idUSKCN1BM0SL.

245 "Under this new regime": Shoshana Zuboff, *The Age of Surveillance Capitalism: The Fight for a Human Future at the New Frontier of Power* (New York: PublicAffairs, 2019), 53.

246 "kick up a fuss": Albert O. Hirschman, *Exit, Voice, and Loyalty: Responses to Decline in Firms, Organizations, and States* (Cambridge, Mass.: Harvard University Press, 1970), 30.

248 "We urgently need more": Quoted in Kate Conger and Daisuke Wakabayashi, "Google Employees Protest Secret Work on Censored Search Engine for China," *New York Times*, Aug. 16, 2018.

248 "to empower employees": Schmidt and Rosenberg, *How Google Works*, 65.

249 "There are serious worldwide": Ryan Gallagher, "Senior Google Researcher Resigns over 'Forfeiture of Our Values' in China," *Intercept*, Sept. 13, 2018, theintercept.com/2018/09/13/google-china-search-engine-employee-resigns/.

252 "We believe that Google should not be": Quoted in Scott Shane and Daisuke Wakabayashi, "'The Business of War': Google Employees Protest Work for the Pentagon," *New York Times*, Aug. 4, 2018, www.nytimes.com/2018/04 /04/technology/google-letter-ceo-pentagon-project.html.

253 "attribute that is deeply discrediting": Erving Goffman, *Stigma: Notes on the Management of Spoiled Identity* (New York: Simon & Schuster, 1963), 3.

253 "Like individuals, organizations": Thomas Roulet, "What Good Is Wall Street?: Institutional Contradiction and the Diffusion of the Stigma over the Finance Industry," *Journal of Business Ethics* 130 (Aug. 2015).

254 "We effectively, if often unthinkingly": Goffman, *Stigma*, 5.

254 "the staff would not end up being a stigmatized group": Bruce G. Link and Jo C. Phelan, "Conceptualizing Stigma," *Annual Review of Sociology*, vol. 27 (2001): 376.

255 "to regard their success as their own doing": Michael J. Sandel, *The Tyranny of Merit: What's Become of the Common Good?* (New York: Farrar, Straus and Giroux, 2020), 25.

256 "Google is using machine learning": Brian Merchant, "How Google, Microsoft, and Big Tech Are Automating the Climate Crisis," *Gizmodo*, Feb. 21, 2019, gizmodo.com/how-google-microsoft-and-big-tech-are-automating-the -1832790799.

258 "control the process of their own work": Ursula M. Franklin, *The Real World of Technology* (Toronto: House of Anansi Press, 1990), 10.

259 "the authority and resources": Annie Kelly, "Apple and Google Named in US Lawsuit over Congolese Child Cobalt Mining Deaths," *Guardian*, Dec. 16, 2019, www.theguardian.com/global-development/2019/dec/16/apple-and -google-named-in-us-lawsuit-over-congolese-child-cobalt-mining-deaths.

260 "Virtually all the companies": "See No Evil, Speak No Evil: Poorly Managed Corruption Risks in the Cobalt Supply Chain," Resource Matters, 2019, 17.

Epilogue

265 "in treating one patient": Jillian Mock, "Psychological Trauma Is the Next Crisis for Coronavirus Health Workers," *Scientific American*, June 1, 2020, www.scientificamerican.com/article/psychological-trauma-is-the-next-crisis -for-coronavirus-health-workers1/.

265 "The collective soul": Nivedita Lakhera, "As a Front-Line Doctor, I Can't Let Another Doctor Suffer Trauma, Suicide," *USA Today*, April 1, 2020, www.usatoday.com/story/opinion/voices/2020/04/01/coronavirus-doctor -colleagues-suffering-trauma-column/5098054002/.

Acknowledgments

I could not have written this book without the support of numerous foundations and fellowship programs. I am particularly grateful to the Dorothy and Lewis B. Cullman Center for Scholars and Writers at the New York Public Library, where I spent a year researching and writing the book. Special thanks to Jean Strouse, its former director; Salvatore Scibona, the current director; and deputy directors Lauren Goldenberg and Paul Delaverdac, whose support and geniality make the Cullman Center such an exceptional place.

I am equally grateful to the Carnegie Corporation of New York, which awarded me an Andrew Carnegie fellowship in 2018, and to Type Media Center, where I have been a Puffin Foundation Writing Fellow. Thanks in particular to Taya Kitman, who has been an invaluable supporter and loyal friend. I also want to thank the Russell Sage Foundation, where I spent several months as a visiting journalist, and in particular Sheldon Danzinger, RSF's president, and Claire Gabriel, who helped me with research. Thanks as well to the New Orleans Center for the Gulf South, which awarded me a Monroe Fellowship to do some research in the Gulf, and to Rebecca Snedeker, its brilliant executive director.

I am immensely indebted to Sarah Chalfant of the Wylie Agency. Sarah is the best agent and advocate a writer could hope to have. Thanks as well to Luke Ingram and Rebecca Nagel for their assistance and encouragement.

No person shaped this book more than Eric Chinski, a brilliant editor who I feel very lucky to have worked with. Eric's passion for ideas, his unerring judgment, and his deep commitment to his authors made me feel accompanied during the long, often lonely process that writing a book can be. Thanks also to Julia Ringo for her excellent editorial suggestions and to Janine Barlow for her expert fine-tuning.

I am grateful to Daniel Zalewski, a legendary editor at *The New Yorker* who encouraged me to investigate the abuses at the Dade Correctional Institution. Working with Daniel has taught me so much about how to craft and report a story. I'm also grateful to Sasha Weiss, with whom I had the pleasure to work on a story about the wounds that encumber drone warriors.

I owe a special debt to Eric Klinenberg, who, a decade ago, encouraged me to apply to the PhD program in sociology at NYU, and who later invited me to become a visiting scholar at the Institute for Public Knowledge. It was through the program that I came across the work of Everett Hughes. I am equally grateful to Steven Lukes, a mentor and intellectual inspiration whose feedback on an early draft was invaluable.

Many thanks to Andy Young for fact-checking the book with care and levity. Thanks also to Margot Olavarrio for helping with translation, and to Sara Feinstein for research assistance.

One thing that makes writing a book less lonely is the support and comradeship of fellow writers and friends. I am particularly grateful to Adam Shatz, Sasha Abramsky, Rowan Ricardo Phillips, Laura Secor, Scott Sherman, Gregory Pardlo, Mona El-Ghobashy, Ari Berman, Steven Dudley, Kirk Semple, Caitlin Zaloom, Chase Madar, Jennifer Turner, Nicole Fleetwood, Neil Gross, and Peter Yost, and to the wonderful group of fellows I got to know at the Cullman Center, in particular Ava Chin, Nellie Hermann, Joan Acocella, Sarah Bridger, Martin Puchner, Blake Gopnik, Hugh Eakin, and Barbara Weinstein.

My deepest debt of all goes to my family: my generous and loving parents, Carla and Shalom; my incredible mother-in-law, Graciela Sas-Abelin Rose, who read a draft of the book and offered valuable feedback; my sister, Sharon, for her love and support; and my brother- and sister-in-law, Laurent Abelin and Suzanne Ehlers, for making the time away from work so memorable and fun. Above all, I am grateful to my wife, Mireille Abelin, whose love enriches my life immeasurably and whose commitment to emotional and intellectual growth challenges and inspires me. She is also my most perceptive and discerning reader, and an amazing mother to our beautiful children, Milena and Octavio.

Index

criminal justice system: mentally ill in, 24, 90–91; sentencing laws in, 24, 43, 53–54, 65, 79–81, 267; shift from rehabilitative ideal to punitive approach in, 53–54; *see also* prisons

Crosby, James V. "Jimmy," 43

Crusius, Patrick, 178*n*

Cummings, Jerry, 28, 30

Curtis, Bill, 49–53, 57, 63–65, 69, 70, 106, 110

Dade Correctional Institution, 57–59, 81, 82, 91, 110, 227

Dade Correctional Institution, Transitional Care Unit (TCU) of, 19–22, 24–30, 34–43, 82, 84–85, 141–42, 246; article on abuse of prisoners at, 30; Corizon's contract with, 84, 85; Cummings as warden at, 28, 30; empty meal trays at, 37; filth and run-down conditions at, 26; guards at, 20–21, 24–29, 35–40, 47–48; Hempstead at, 29–30, 38–39, 47; Krzykowski at, 19–22, 24–30, 37–39, 44, 47, 48, 84, 85, 110, 141–42, 246, 249, 269–70; lawsuit against Florida DOC for abuse at, 40, 94; Mair at, 64; Mallinckrodt at, 25, 27–28, 36–37, 41–42, 81, 85, 141, 142, 246; Morris as assistant warden at, 40–42, 84–85; Perez at, 19–21, 25, 41; Rainey's murder at, 26–27, 29–30, 34, 35, 38, 39, 42–44, 47–48, 64, 70, 71, 81, 82, 84, 91–96, 141, 246; recreation yard access at, 20; Richardson at, 35–36; shower abuse at, 27, 29, 39, 47–48, 82, 110, 246; solitary confinement at, 21, 42; Wexford's contract with, 84–85

data collection, 257

Datta Khel, 151

Dauphin Island, 221

Dean, Jeff, 248

Dean, Wendy, 265

Deepwater Horizon, 222; Horizon oil spill, 211–13, 216–21, 224, 230–38

Defense, U.S. Department of, 101; Project Maven with Google, 251–52, 256, 257

Defense Production Act, 195

Deitch, Michele, 83

Dell, 259, 260

democracy, 134

democrats, passive, 13–14, 268

Desert Rock Airport, 149

Desert Waters Correctional Outreach, 61

Devereaux, Ryan, 151

Diagnostic and Statistical Manual of Mental Disorders, 115

Dickens, Charles, 76, 78

dirt, 179

Dirty Jobs, 125

dirty work: automation of, 256–57; essential features of, 11–12, 252–53, 266; "good people" and, 6, 7, 12–13, 256, 267; as hidden and isolated, 10, 12–13; Hughes's essay on, 5–7, 14, 48–49, 88; inevitability of, 11; invisible contract in, 15; least resistant personalities and, 224, 228; passive democrats and, 13–14, 268; scandals and, 10, 269

dirty work, use of term: familiar meaning, 11; by Hughes, 5; in this book, 11–12

dirty workers, 7, 9–10; moral burden of, 8–11

disability benefits, 23

virtuous consumption and, 207–208; wages at, 162, 166–67; white workers at, 169–70, 181

slavery, 66–67, 170

slave traders, 66–67

Sleasman, Peter, 93–94

smartphone batteries, cobalt for, 258–63, 269

Smith, Michael Patrick F., 232

Smithfield Foods, 196, 201, 204–206

Snowden, Edward, 147

social control, 34

social media, 245, 258

Social Problems, 4–6

Social Relations in Our Southern States (Hundley), 66

Society of Captives, The (Sykes), 55

Soleimani, Qassem, 104

solitary confinement, 21, 42, 78, 91–92

Solnit, Rebecca, 151

Somalia, 104, 105, 129

Some Thoughts Concerning Education (Locke), 164

Soul by Soul (Johnson), 67

Southern Poverty Law Center, 86–87

Soviet Union, 243

Spiegel, Der, 243

Spinaris, Caterina, 60–61

spoiled identity, 253–55

Stanford International Human Rights and Conflict Resolution Clinic, 151

Stanford University, 247

stigma, 253–55, 268

Stigma (Goffman), 253–54

Stone, Sara Lattis, 211–12, 215–18, 221, 233–40; filmmaking of, 235, 237–39; paintings of, 215–17, 235, 238–40

Stone, Stephen, 211–15, 217–23, 233–36, 238–40, 256, 267

Stuesse, Angela, 169–70, 199

Sullivan, Kenneth, 205

Supreme Court, U.S.: *Estelle v. Gamble*, 22–24; *O'Connor v. Donaldson*, 23

surveillance, 147; Project Maven, 251–52, 256, 257

Sweatt, Loren, 196

Sykes, Gresham M., 55

Taliban, 113, 114, 135, 136

Tannenbaum, Frank, 54–55

targeted assassinations, 103; *see also* drone program

Teamsters, 64

Tea Party, 224

technologies, holistic versus prescriptive, 258

tech sector, 241–63; Apple, 258–61; cobalt mining for batteries in, 258–63, 269; corruption in, 260; diffusion of responsibility in, 261; Google, *see* Google; nondisclosure agreements in, 262, 269; supply chains in, 258, 261–62

10-20-Life law, 65, 80–81

terrorism: September 11 attacks, 100, 104, 109, 128, 134, 149, 187; war on, 101, 103, 105–6, 112, 113; *see also* drone program

Texas, 22, 173; Bryan, 158, 161, 168–70, 172–77, 181, 194, 197, 199; migrant farmworkers in, 178–79; "shadow people" in, 157–58, 167–68, 178

Texas Education Agency, 173–74

Theriot, Jason, 225–26

Things They Carried, The (O'Brien), 115

This American Life, 161–62

Thompson, Chris, 79, 80

Thompson, Judy, 78–84

A Note About the Author

Eyal Press is an author and a journalist based in New York. The recipient of the James Aronson Award for Social Justice Journalism, an Andrew Carnegie fellowship, a Cullman Center fellowship at the New York Public Library, and a Puffin Foundation fellowship at Type Media Center, he is a contributor to *The New Yorker*, *The New York Times*, and numerous other publications. He is the author of *Beautiful Souls* and *Absolute Convictions*.